Lunar Science

Edited by Yann H. Chemin

Published in London, United Kingdom

IntechOpen

Supporting open minds since 2005

Lunar Science
http://dx.doi.org/10.5772/intechopen.78817
Edited by Yann H. Chemin

Contributors
Robert J. Reynolds, Yuriy Khachay, Boris P. Kondratyev, Bryan Palaszewski, Abdulrahman Malawi, Olga Hachay, Oleg Khachay, Yann H. H. Chemin

First published in London, United Kingdom, 2019 by IntechOpen
IntechOpen is the global imprint of INTECHOPEN LIMITED, registered in England and Wales, registration number: 11086078, The Shard, 25th floor, 32 London Bridge Street
London, SE19SG – United Kingdom
Printed in Croatia

British Library Cataloguing-in-Publication Data
A catalogue record for this book is available from the British Library

Additional hard and PDF copies can be obtained from orders@intechopen.com

Lunar Science
Edited by Yann H. Chemin
p. cm.
Print ISBN 978-1-78984-127-5
Online ISBN 978-1-78984-128-2
eBook (PDF) ISBN 978-1-83881-854-8

For EU product safety concerns:
IN TECH d.o.o., Prolaz Marije Krucifikse Kozulić 3, 51000 Rijeka, Croatia, info@intechopen.com or visit our website at intechopen.com.

Meet the editor

Dr. Yann H. Chemin's research spans geography, cartography, energy balance, agriculture, land and water processes, and more recently rocky (minor) planets and the moon. His credentials are a BSc in International Agri-Development (1995, France), an MSc in Land and Water Resources Management (1996, UK), a DTSc in RS/GIS Applications (2006, Thailand), and a BSc in Planetary Sciences with Astronomy (2016, UK). He worked on the Chandrayaan-1 M3 sensor and its hyperspectral applications on the Apollo 12 landing site. Most recently, he also worked on Ceres craters. As of May 16, 2019, he is joining the JRC, Italy.

Contents

Preface

Lunar science is in the middle of a small revolution, with now many new countries sending orbiters, landers, and even sample return missions to the moon. Additionally, both governments and private companies are now more and more considering the moon as a base for solar system exploration. With such an increase in attention, lunar science is now encompassing several unified dimensions.

The first is reflected in the section "Early and Inner Moon," which is the science of the moon itself, its initial evolution, its mass center deviation, and methodologies for assessing deformation processes in its deeper layers.

The second is discussed in the section "Lunar Surface and Humans," which looks at how humans can live on the moon, a certainly different perspective to common lunar science. This is a more practical direction that fits a single goal: facilitating or simply enabling the survival of human beings on the surface of the moon for a time period superior to the short-term visits of the Apollo era. To reach that goal, scientific fields of study covering biology and in general the logistics of surviving there are counted, including in-situ resource utilization and the more traditional field of surface mapping.

Yann H. Chemin
JRC, Italy

Section 1

Introduction

Chapter 1

Introductory Chapter: A Tipping Point for a Return to the Moon

Yann H. Chemin

1. A new economic influx towards a cheaper access to the Moon

We are living at the onset of another return to the Moon, this time during another phase of our planetary development, a more peaceful one. With the rise of private access to space (**Table 1**), the cost of sending cargo/crew to Low Earth Orbit (LEO) is plummeting, and to add to it, with a rocket relaunch timetable of 15 days. It is a simple corollary to think that the short-term assembly of a large spacecraft in LEO can be done. As a consequence, humanity could reach to the Moon cheaper with logistical support for much longer missions than what was done 50 years ago, during the golden age of Luna and Apollo scientific missions.

Waiting for the wake of the giants and the first manned missions to the Moon (again, at least in this twenty-first century) and through to Mars (**Figure 1**), the Moon resources have had a new rekindled interest by researchers within the last 20 years with the Clementine and Chandrayaan missions, among others. The Moon, over these two last decades, has become the exploratory ground to missions from different countries, some of them new to space probe building: Israel (Beresheet), India (Chandrayaan-1/Chandrayaan-2), China (Chang'e-1–Chang'e-4), Japan (Kaguya) and European Union (SMART-1). Some of the missions have a lander component (Beresheet, Chandrayaan-2, Chang'e-3/Chang'e-4), and even some have a mobile robotic explorer (Chang'e-3). Surface/elevation mapping, exosphere, radiation and volatile composition are the main subjects of interest for mission controls around the world, besides the engineering proof of work for orbital control, landing automation and sample return. Already China is launching a sample on the far side of the Moon this year (Chang'e-5), India being both an orbiter and a lander (Chandrayaan-2). In 2020, Russia is planning a lander (Luna-25). Finally, in 2021, Japan will also land with its SLIM mission. The number of missions including landing components is steady in these coming years; however, the only known manned mission towards the Moon is EM-2 orbital mission of the USA, planned

Company	Cargo to LEO	Crew to LEO
Blue Origin		Development
Boeing		Development
Northrop Grumman Innovation Systems	Operational	
Orbital	Retired	
Sierra Nevada Corporation	Development	Development
SpaceX	Operational	Operational

Table 1.
Private transport to LEO (operational/development).

IntechOpen

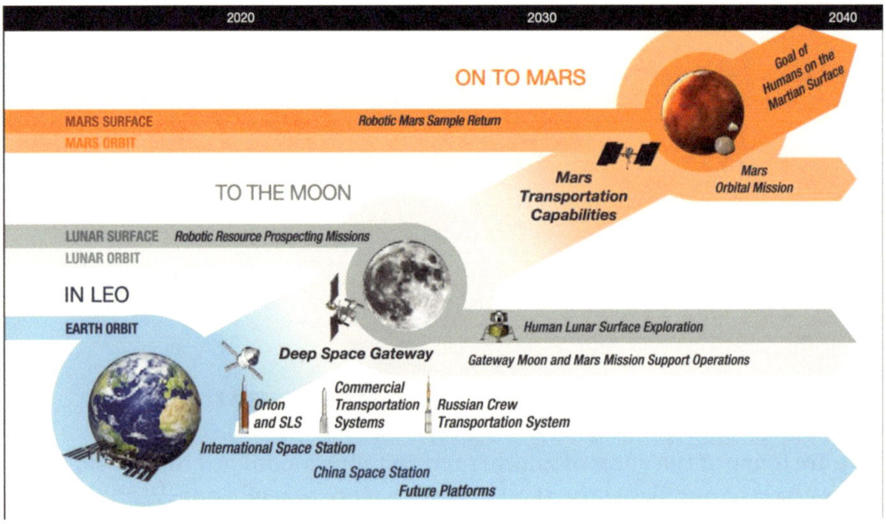

Figure 1.
The Global Exploration Roadmap [1].

Company	Company
Asteroid Mining Corporation Ltd., UK	NEO Resource Atlas
Aten Engineering	OffWorld
Deep Space Industries	Planetary Resources
Kleos Space	Planetoid Mines Company
Moon Express	TransAstra

Table 2.
Asteroid mining companies.

in 2023. This leaves humans on the Moon to another decade probably, unless some country/company wants to make a statement by bringing the element of surprise and precipitating this prospect.

The Moon has several interesting conditions that humanity can use. It has an enhanced slingshot effect to the Earth, coupled with a much lower gravity. Launching into the solar system from the Moon has significant advantages. One of them could develop with the asteroid mining/capture private endeavours (**Table 2**). Its lack of atmosphere also brings advantages to astronomical observation, more so with a large distributed array [2].

2. Humanity near-future: studying the Moon on-site

The prospect of such various interests (and others) having the Moon in the midst of their plans brings also the possibility of surface dwellers. This translates into a new interest in cheaper field survey from surface dwellers [3] giving a vastly deeper geological understanding of the lunar surface elements/processes. This also opens the way for installing seismologic and heat transfer monitoring stations for a more automated monitoring of the Moon interior. Similarly, human health studies will certainly make major strides from the first longer periods on the Moon surface [4, 5].

From water to ^{3}Helium [6], various types of useful resources can be extracted and activated if appropriate machinery can be transported on-site or various processes developed to extract actively/passively useful elements from the regolith. Within that perspective, this already drove the choice of landing the China rover on thorium-rich regolith. It turns out that In Situ Resource Utilisation (ISRU) becomes paramount to improve the residence conditions of humans living (medium or long term) on the Moon.

It may take another 20 years to see some practical set-up of humans actually being on the Moon, taking two generations to reach back to the lunar surface and picking up field research where we prematurely stopped it in the 1970s, with a 70-year recess. In the meantime, it becomes rather obvious that the Moon is renewed in its aura of the Earth's first harbour to the solar system and the first non-Earth scientific playground.

Author details

Yann H. Chemin
JRC, Italy

*Address all correspondence to: dr.yann.chemin@gmail.com

References

[1] ISECG. The Global Exploration Roadmap, January 2018. The International Space Exploration Coordination Group; 2018. https://www.globalspaceexploration.org/

[2] Crawford IA, Anand M, Cockell CS, Falcke H, Green DA, Jaumann R, et al. Back to the Moon: The scientific rationale for resuming lunar surface exploration. Planetary and Space Science. 2012;**74**(1):3-14. https://arxiv.org/abs/1206.0749

[3] Crawford IA. Science enabled by a Moon Village. Earth and Planetary Astrophysics. 2017. arXiv:1706.06698. https://arxiv.org/abs/1706.06698

[4] Khan-Mayberry N. The Lunar Environment: Determining the Health Effects of Exposure to Moon Dusts. PhD Dissertation. Houston, TX: NASA; 2007. https://ntrs.nasa.gov/archive/nasa/casi.ntrs.nasa.gov/20070006527.pdf

[5] Linnarsson D, Carpenter J, Fubini B, Gerde P, Karlsson LL, Loftus DJ, et al. Toxicity of lunar dust. Planetary and Space Science. 2012;**74**(1):57-71. https://arxiv.org/abs/1206.6328

[6] Crawford IA. Lunar Resources: A Review. Progress in Physical Geography. 2015;**39**(2):137-167. https://arxiv.org/abs/1410.6865

Section 2

Early and Inner Moon

Chapter 2

Initial Evolution of the Moon

Khachay Yuriy

Abstract

The problem of the origin of the Moon is of fundamental importance to understanding the mechanism of the planetary solar system's formation. It is important to know the mechanism of differentiation of substances in a growing planet. When planets are formed from a cold protoplanetary cloud, the matter of the inner regions of the Earth and the Moon remains at temperatures lower than the melting point of iron. The main volume of the matter of the protoplanet remains in its unmelted state, and its differentiation occurs in the formed planet. In this work, attention is paid to the most important internal sources of energy: the decay energy of short-lived isotopes, the dissipation of tidal friction energy, and thermal energy from accidental deposition of bodies and particles on a growing surface. Accounting for these sources provides a solution to the problem.

Keywords: moon matter, composition, age, growth dynamics, numerical models

1. Introduction

The study of the evolution of the moon is of exceptional interest for the knowledge of the processes which had occurred and occurred now on the Moon and the Earth, the formation of the internal regions of these bodies. Of undoubted interest is the study of the material composition of the surface of the planetary body nearest to the Earth, which makes it possible to clarify information about the processes in the protoplanetary cloud and the early stages of the accumulation of planets.

Before to the work of descent vehicles on the surface of Mars, the only direct information about the composition of the substance of external bodies was the results of a study of lunar soil, delivered to Earth by Soviet stations Luna-16 and Luna-20 and American astronauts. The geophysical measurements of the gravitational potential and the speed of propagation of seismic waves made it possible to build indirect models of the distribution of the density and velocity sections of the Moon [1]. For the no uniqueness of the geophysical interpretation, estimates of the density distribution and velocity of seismic waves for 1D models of the Moon that are very important for the knowledge of the current state are obtained. The lunar crust is fixed from the surface to a depth of about 55 km. The velocity of longitudinal seismic waves V_P increases here to a value of V_P = 5.8 km/s at a depth of 25 km, then sharply increases at this depth to a value of 6.8 km/s. Then the velocity increases slowly down to 7 km/s up to the 55 km border. This section is called the crust-mantle transition boundary. The surface is marked by a sharp increase in the velocity of V_P to the value of V_P = 8.1 km/s. Then there is a decrease in the velocities V_P and Vs to values of 7.8 km/s and 4.7 km/s, respectively, at a depth of 300 m. The layer in the depth interval (55–300) in km is called the upper mantle. In the range of

300–800 km, called the average mantle, the velocity is reduced to V_P = 7.5 km/s and V_S = (3.6–4.0) km/s.

In the lower part of the mantle, transverse waves are not recorded. At a depth of about 1500 km, the mantle-core transition region is fixed, and in the second one, there is a sharp decrease in the propagation velocity of the volume seismic waves to V_P = (4–5) km/s [2–5]. In spite of the no uniqueness of the geophysical interpretation, very important results of the density distribution and velocity of seismic waves for 1D models of the Moon had been obtained. The scheme of the seismic velocity model is presented in [1]. The nature of the propagation of seismic waves is very different from that observed in the Earth: the amplitude of oscillations increases sharply, and the decline is observed for (1–4) hours. The lunar crust differs significantly from the earth's crust in its elastic and viscous characteristics. The quality factor (the inverse of the attenuation coefficient) is estimated at 5000, while for the Earth, this estimate is in the range of 100–1000. The seismic activity of the moon is much smaller, about 1015 erg/year, whereas on Earth about 1024 erg/year.

Based on the mineralogical study of the delivered collection of lunar samples, it was established that, unlike the Earth, an early and extensive differentiation of matter took place on the Moon. At the same time, at the early stage of the formation of the Moon, the fractional crystallization of the substance, which formed, possibly, the entire Moon, followed by partial melting of the upper envelope, with a capacity of at least 250 km, occurred. Lunar magma was formed during the entire time of its cooling [6]. The question of the origin and composition of the substance of the moon is of fundamental importance for understanding the processes of formation of the planets of the solar system. Before the occurring of the mega-impact hypothesis, three main mechanisms for the formation of the Moon had been discussed in the literature: (1) the hypothesis of separation of the moon from the earth, (2) the capture hypothesis, and (3) hypothesis of the co-formation or co-accretion of the Earth and the Moon. The disadvantages of these hypotheses are considered in [7]. The idea of separating the substance of the Moon from the Earth was proposed by Darwin in 1880 [8]. Its inconsistency with the laws of celestial mechanics is considered in [9]. As the authors of [9] note, in the event of a rotational instability, which causes the separation of a part of a substance from a rotating body, a smooth separation of a satellite from the main body is impossible. The substance ejected as a result of rotational instability either flies away or returns back [9].

Later Ringwood [10] attempted to modify Darwin's hypothesis by assuming that the material from the Earth's mantle was ejected into the Moon's orbit by strikes of large meteorites. The hypothesis of a joint formation of the Earth and the Moon is considered in [11]. Schmidt [11] assumed that the Moon had accumulated in the vicinity of the growing Earth from a near-Earth swarm of bodies, which was continuously increased from a protoplanetary cloud. As the authors of [12] note, "the hypothesis of O. Yu. Schmidt is based on processes that necessary must take place during the accumulation of the Earth and from the mechanical point of view it seems to be the most promising [12]." However, within the framework of this model, it was not possible to explain the difference in the chemical composition of the Earth and the Moon. The main problem that arose in connection with the thermal state of the Moon is the need to substantiate the source of the early differentiation of matter at the stage of their accumulation. To overcome the difficulties of the three main hypotheses of lunar formation, the mega-impact hypothesis [13, 14] was proposed.

In our work, the task was set to construct a numerical model of the Moon accumulation process, in which we would be able to reconcile the experimental obtained data. Taylor [15] believes that if the Earth-Moon system is unique, it is possible that its genesis is unusual and this unusual variant is the hypothesis of the formation of the Moon as a result of the mega impact, the Earth's collision with a space body of planetary size (with mass of Mars or more). A number of geochemical contradictions that are incompatible with the mega-impact hypothesis are considered by Galimov in [7].

We were faced with the task: using the model of accumulation of the terrestrial planets proposed in papers [16–18], to conduct numerical simulations of the temperature distribution in the interior of the planet for successively increasing with time values of the body radius in the 3D environment and in contrast to the results [18], explore the features of the evolution of primary heterogeneities depending on the rate of accumulation of the second largest body in the "feeding" zone of a growing Earth in a 3D process model. In works [16–18], the proposed model of the heterogeneous accumulation of the Earth, based on a two-stage mechanism for the formation of the pre-planetary of the planet, is presented. It assumes that in the first stage, primary pre-planets are formed, the central part of which consists of the most high-temperature condensates close in composition to the CAI—Ca-Al inclusions found in the Allende meteorite. The middle envelope of these pre-planets is represented by an iron-nickel material, which condenses from the gaseous phase following high-aluminum condensates. In the process of growth of primary pre-planets, they were heated, as a result of the decay of short-lived radioactive isotopes, the main of which is ^{26}Al with a half decay of $\tau = 7.38 \times 10^5$ years [19]. The ratio of $^{26}Al/^{27}Al$ in protoplanetary matter is estimated to be 5×10^{-5} [20]. With this content of ^{26}Al, as the mass of the pre-planetary grows, the temperature of their central regions increases, and in the center of the pre-planetary with a radius of more than 200 km, it can reach 2200 K [17]. This is quite enough to melt the Ca-Al material in the central part of the core, the melting point of which is 1830 K [1], and the iron-nickel mixture in its middle envelope. The outer envelope of the pre-planetary, which transfers heat to the space, will remain solid. Further development of the process of formation of the planets is as follows. In accordance with the accumulation model of Safronov [21], the number of cores formed at the initial stage of the agglomeration process of condensation products is large, and they will often collide with each other. The collision of two pre-planetary with similar sizes, molten aluminosilicate core and middle envelope, folded with iron and a solid silicate outer envelope will lead to their destruction. Medium, molten envelopes in a collision will merge, forming a new pre-planet, the core of which consists of an iron-nickel alloy. The substance of the aluminosilicate core of the primary core will be extruded from their centers and thrown into the feeding zone for new core formed as a result of the collision. The outer hard envelope, the lower part of which could consist of a substance close to the composition of pallasites or ordinary chondrites, will be destroyed, and part of the fragments will also be thrown out of the feeding area of the growing planet. In this way, the metallic core of the Earth and the Moon is formed, and the separation of their chemical reservoirs of the core and the mantle takes place.

2. Mathematical modeling of the temperature distribution in the moon at the stage of its accumulation

The model [21] on the formation of planets and their satellites from a protoplanetary cloud is used as the initial one. The results of numerical modeling obtained by us [16, 17] showed that already at an early stage of the Earth's accumulation process, heat release during decay by short-lived naturally radioactive elements and above all ^{26}Al is enough to exceed in the protoplanetary dimensions (50–100 km); a molten central region and a relatively thin, solid, predominantly silicate upper envelope could be formed. This creates the conditions for the mechanism of differentiation of a substance into a chemical reservoir of the future predominantly iron core and a silicate reservoir of the Earth's mantle to be realized at small relative velocities of collisions of small bodies—pre-planets. The molten parts were combined, and the solid, mostly silicate, could not yet be held by the weak gravitational field of the planet and returned to the "feed zone." It is this material component that provided the chemical reservoir, from which the second largest body of the Earth-Moon system was formed. Fractionation went on in small bodies and was mostly completed in less than 10 million years, while the formation of the structure of the core and the mantle of the Earth continued for about 100 million years. The union of the liquid internal parts of the colliding bodies occurred as a result of inelastic collision; therefore, most of the potential gravitational energy was converted into heat through the kinetic energy of the collision. The process continued until the core reached a large part of the modern mass. At the final stage of the growth of the pre-planet, the mass of the Earth's pre-planet is already sufficient to hold an ever-increasing proportion of the silicate envelope of the falling out bodies. And the composition of the growing area is increasingly enriched with an admixture of silicates. The process of a completely inelastic collision of accumulated bodies with a high degree of potential energy conversion by gravitational interaction into heat, gradually turned into a mechanism of almost solid-state collisions, in which only a small part of the kinetic energy is converted into heat absorbed by the planet's pre-planet.

The carried out mathematical modeling of the thermal evolution of a growing planet implements the described process scheme. For the growth rate of the pre-planet of the planet, the Safronov model is used in the variant [21]:

$$\frac{\partial m}{\partial t} = 2(1+2\theta)r^2\omega\left(1-\frac{m}{M}\right)\sigma \tag{1}$$

where ω is the angular velocity of the orbital motion, σ is the surface density of the substance in the "feeding" zone of the planet, M is the modern mass of the planet, r is the radius of the growing pre-planet, and θ is a statistical parameter taking into account the distribution of particles by mass and velocity in the "feeding" zone. Usually, the study of changes in PT conditions in growing satellites of the planets is not paid attention. In this work, on the contrary, we aim to study how strongly these conditions can depend on the rate of increase in body weight. Let us use the estimate for the mass of the largest-growing body m and the next largest body m_1 obtained in the paper [12, 22] for the parameters of bodies in the Earth's group:

$$\frac{m_1}{m} \approx \frac{1}{3}\left(1-0.6\left(\frac{m}{M}\right)^{0.3}\right) \tag{2}$$

The mathematical description of mass-energy transport in a growing self-gravitating body of variable radius consists in setting boundary value problems for a system of equations for the balance of momentum, energy, conservation of mass of matter, and the Stefan problem at the boundaries of regions with melt zones [16–18]:

$$\rho\left[\frac{\partial \vec{V}}{\partial t}+\left(\vec{V}\,\nabla\right)\vec{V}\right] = -\nabla P+\eta\Delta\vec{V}+\left(\frac{\eta}{3}+\xi\right)\nabla\left(\nabla\,\vec{V}\right)-\rho\nabla W \tag{3}$$

$$\rho T\left[\frac{\partial S}{\partial t}+\left(\vec{V}\,\nabla\right)S\right]=\lambda\Delta T+Q \tag{4}$$

$$\Delta W_1=-4\pi\gamma\rho \quad W=W_1+W_2 \tag{5}$$

$$\frac{\partial\rho}{\partial t}+\nabla\left(\rho\,\vec{V}\right)=0 \tag{6}$$

$$L\frac{\partial\,\vec{\psi}}{\partial t}=\vec{q}\Big|_{\xi+0}-\vec{q}\Big|_{\xi-0} \tag{7}$$

where $\vec{V}$ is fluid velocity, P is pressure, S is entropy, W_1 is gravity potential, W_2 is centrifugal potential, ρ is density, η and ξ are coefficients of the first and second viscosity, λ is thermal conductivity coefficient, γ is gravitational constant, Q is total the power of internal energy sources per unit volume, L is the heat of phase transition, $\frac{\partial\vec{\psi}}{\partial t}$ is the velocity of displacement of the interface, $\vec{q}\Big|_{\xi+0}$ *and* $\vec{q}\Big|_{\xi-0}$ the heat flux density, respectively, before and after the phase boundary, and ∇ and Δ are the operators Nabla and Laplace.

The main difficulties are connected with the solution of the boundary value problem for the Navier-Stokes equation (Eq. 2). Even in the approximation with constant viscosity coefficients, as used in [16], finding a numerical solution in a 3D spherical layer is a significant problem. In addition, within the framework of equation (Eq. 2), it is difficult to describe the forced convective mixing of a substance near the surface of a growing body when individual bodies fall. The temperature distribution in the body of increasing radius is found from the numerical solution of the boundary value problem for the heat equation, taking into account the possibility of a melt appearing without explicitly highlighting the position of the crystallization front and parametric accounting for convective heat transfer in the melt [23]:

$$c_{ef}\rho\frac{\partial T}{\partial t}=\nabla\left(\lambda_{ef}\nabla T\right)+Q \tag{8}$$

where c_{ef}, λ_{ef} is the effective values of heat capacity and thermal conductivities, which take into account the heat of melting in Stefan's [24] problem and the presence of convective heat transfer, T is the sought temperature at a point at time t, and Q is the volume power of internal heat sources. The problem was solved by the finite difference method using an implicit, monotonic, conservative scheme. In Eqs. (1)–(8), the steps on the temporal and spatial grids are the same. The time grid step is variable and, with the density distribution chosen, as a function of depth, is calculated from Eqs. (1)–(2). Using these equations, at each time step, the mass of the growing planet and the distribution of lithostatic pressure in internal regions are calculated. For each value of the achieved size of a growing planet, the melting temperature distribution is calculated. In the core, the dependence of the melting point of mainly iron composition is calculated according to [25]. In the region of the predominantly forming silicate mantle, the dependence of the melting point on pressure is used [26]. The zone of complete and partial melting was determined for each time layer by comparing the calculated temperature distribution with the distribution of the melting temperature at a given depth.

Conditions are set on the surface of the pre-planet to ensure the balance of the incoming part of the potential energy of the gravitational interaction of bodies, the expenses of heat for heating the incoming substance and the heat flow re-emitted into the space taking into account the transparency of the external environment:

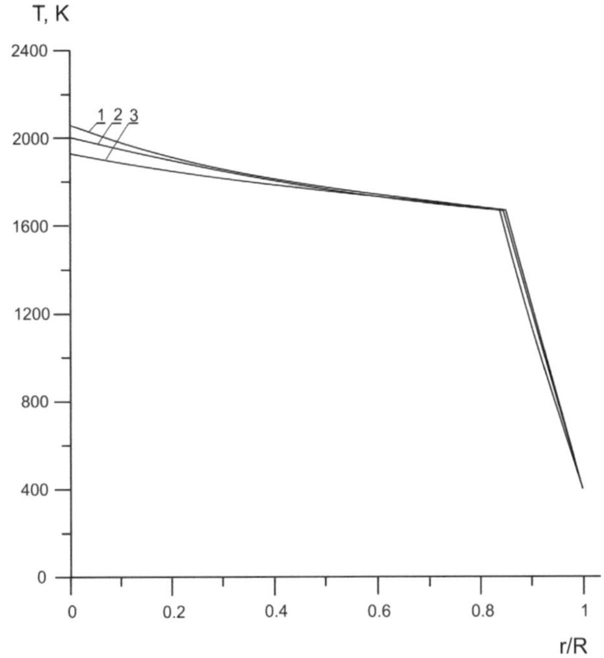

Figure 1.
The distribution of the temperature in the model of pre-planetary body. Its radius is (1) 400 km, (2) 300 km, and (3) 250 km [16].

$$k\rho \frac{\gamma M}{r}\frac{dr}{dt} = \varepsilon\sigma\left[T^4 - T_1^4\right] + \rho c_P[T - T_1]\frac{dr}{dt} \quad (9)$$

where ρ is the density of matter, γ is the gravitational constant, M is the mass of the growing planet, and r is its radius. T and T_1 are, respectively, the body temperature at the boundary, the external environment ε is the coefficient of transparency of the medium, c_p is the specific heat, k is the fraction of the potential energy converted to heat, and σ is the Stefan-Boltzmann constant. In Eq. (9), just as in Eqs. (1)–(8), the steps on the temporal and spatial grids are used the same.

The qualitative difference between the obtained variants of the results of numerical simulation in a 3D model is that it was possible to trace the occurrence of thermal and density heterogeneities. The occurrence of these heterogeneities is due to the random distribution of bodies and particles by mass and velocity, which is described by parameter θ in Eq. (1) and is taken into account in Eq. (9) using a random number generator that determines M when calculating the left side of this equation inside the layer, on which is an increase in the radius of the body over time. On **Figure 1** we show the results of calculating the temperature for a one-dimensional spatial model of a growing body.

3. Results and discussion

The results of the numerical solution of the problem for the 3D model of the environment [16, 17] are obtained, which allow us to trace the formation and further dynamics of three-dimensional anomalous in temperature and composition areas resulting from the fall of bodies and particles on the surface of a growing planet when they are randomly distributed over masses. Quantitative estimates of the parameters that determine the solution of the problem are extremely difficult.

One can only hope that further mineralogical and geophysical research results will reduce this uncertainty. One of the results obtained for the temperature distribution over the cross section of the globular sector of the growing Moon model is shown in **Figure 2a**. As can be seen from the above results, with the rapid growth of the Moon, the concentration of short-lived radioactive elements in the center of the growing body is significant, and their contribution to the energy balance can provide temperatures of 2500–3000 K in the region of R < 300 km. For large values of the lunar radius, numerous melt inclusions are recorded, which, with further evolution of the body, tend to unite. Thus, it becomes possible to trace the formation of the "ocean of magma" [27], the presence of which, according to modern concepts, is necessary for the formation of a powerful lunar crust of an anorthosite composition. **Figure 2b** presents the numerical results of the possible temperature distribution for a hypothetical body, the rate of increase in mass which is set significantly less compared to the rate of increase in the mass of the moon when it accumulates at (Eq. 1). This simulates the accumulation of the next mass in the Earth-Moon system. For this option, a slow increase in the volume of the nucleus leads to the fact that most of the decay energy of short-lived radioactive elements has time to dissipate into space and the temperature of the central region barely exceeds the melting point of pure iron. For large values of the radius in the variation of the rate of growth of body weight for the conditions of **Figure 2b**, the distribution of the current temperature values is lower than the melting temperature at a given lithostatic pressure. This eliminates the possibility of endogenous formation of a powerful anorthosite lunar crust and thereby imposes restrictions on the models of the formation of the moon. Common to the physical conditions considered in the variants of **Figure 2a** and **b** is that, due to the small magnitude of the gravitational acceleration, convective heat and mass transfer at the considered depths remains weak and for the analyzed period of time, the randomly distributed thermal heterogeneities caused by the falling of bodies still locally persist. Their presence predetermines the further dynamics of the bowels of the growing Moon.

It is interesting to compare the resulting numerical modeling of the structure of the lunar crust, mantle, and core with the crust, mantle, and core at the stage of the Earth's accumulation. The distributions of hydrostatically varying pressures, as well as melting points in these structures, lead to their qualitative change. Thus, the relative

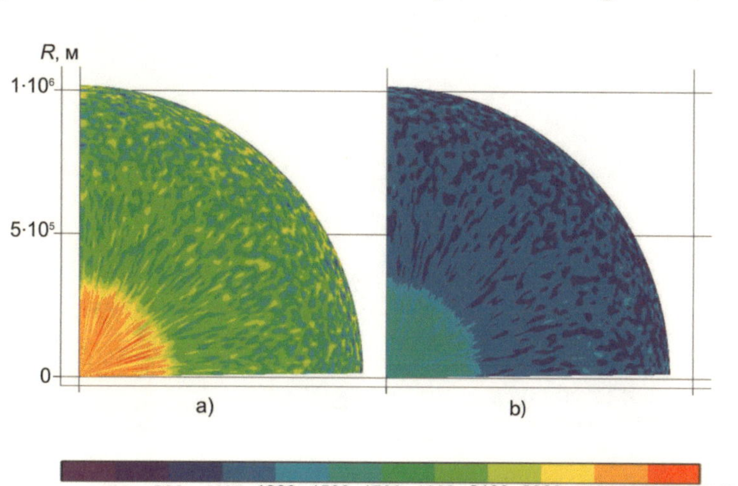

Figure 2.
The dependence of temperature distribution in the three-dimensional sector of the growing moon on the growth rate of its mass by the time the radius reaches R = 1000 km. (a) a model with the physical parameters of the moon; (b) slow growth rate by the ratio (Eq. 2).

proportion of the lunar mantle, as can be seen from **Figure 2a** and **b**, is much larger than the fraction of the Earth's mantle (**Figures 3** and **4**) [18]. The emerging boundaries in the process of accumulation of the crust-mantle of the Moon and the crust-mantle of the Earth are much more irregular, which is apparently due to significantly different pressure values at their boundaries. The same can be noted for the boundaries of the core-mantle of the Moon and the Earth. For the Earth, there is a vast solid core and an external molten core, whereas for the Moon, the inner core is either completely absent or does not appear significantly. In the future, already at the geological stage of development, this may lead to a significant difference in the mineral composition of the moon's crust from the crust of the earth. The lunar crust can be predominantly or even exclusively basalt composition.

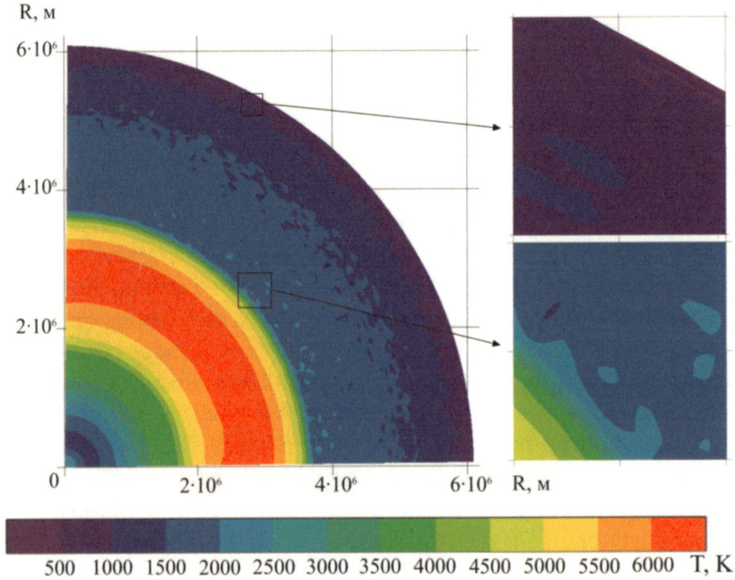

Figure 3.
The temperature distribution of the Earth's interior to the end of accumulation along the section of the 3D sector (without taking into account the heat dissipation energy of tidal friction) [18].

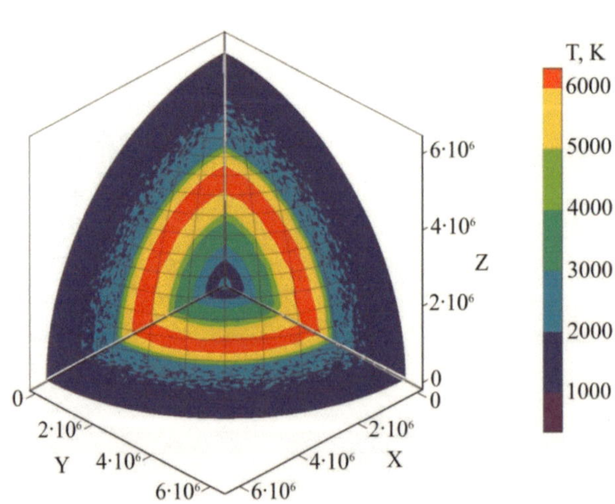

Figure 4.
An example of the temperature distribution and initial thermal heterogeneities in the protoplanet up to the end of its accumulation, without taking into account the heat of tidal friction [18].

4. Conclusions

A numerical solution of the problem is obtained for successively varying temperature and mass distributions over the cross sections of the three-dimensional spherical sector of the Moon model at the stage of its accumulation. It is shown that solutions can be obtained for temperature distribution, based on modern estimates of the physical parameters of the Moon, which provide the endogenous origin of a powerful anorthosite lunar crust. The random distribution of heterogeneities in the inner parts and on the surface of the Moon, caused by the fall of bodies on the growing surface during the accumulation process, controls the initial conditions of the planet's dynamics.

The initial conditions for the Moon and the Earth are taken for a body with a radius of 10 km, the average composition of which corresponds to the matter of carbonaceous chondrites [17] from the common feeding zone of the Earth and the Moon. With their further growth in numerical simulation on the 3D surface of the sphere and at each step of the time grid, which means the radius value, the changing boundary conditions are calculated in accordance with Eq. (5) [18]. These conditions reflect the random distribution of the accumulated bodies in size, composition, and velocity of impact with both the growing Moon and the Earth, which leads to different composition and structure of their internal regions (**Figures 2** and **3**).

In the results presented in **Figure 2a** and **b**, in the initial conditions, the concentration of Al_2O_3 is different: **Figure 2a** reflects the increased content of Al_2O_3, and **Figure 2b** shows the reduced content of Al_2O_3 [17]. To determine the preferred numerical simulation variant, additional geochemical and geophysical space–time data are required.

Acknowledgements

The author thanks his colleague Antipin A.N. for the participation in obtaining solutions to the problem.

Author details

Khachay Yuriy
Institute of Geophysics UB RAS, Yekaterinburg, Russian Federation

*Address all correspondence to: yu-khachay@yandex.ru

References

[1] Sagitov MG. Moon's gravimetry M. Science. 1979. 432 p

[2] Kovach RL,Watkins JS, Talwani P. Moon. The velocity structure of the lunar crust. Moon. 1973;**7**(1–2):63-75

[3] Latham C, Ewing M, Dorman J, Press F. All seismic data on the moon. Science. 1970;**170**(3958):620-626

[4] Nakamura Y, Lammlein D, Latham G. Science. 1973;**181**(4094):49-51

[5] Toksoz M, Press F, Dainty A, et al. Velocity structure and proportion of the lunar crust. Moon. 1972;**4**(3-4):490-504

[6] Frondel J. Lunar mineralogy. New York and London: J. Willey & Sons; 1975

[7] Galimov EM. Dynamic model of the moon formation. Geochemistry. 2005; (11):1139-1150

[8] Darwin GH. Satellite planet revolving around a tidally distributed planet. Philosophical Transactions. Royal Society of London. 1880;**171**: 713-891

[9] Ruskol E. Origin of the Moon. Moscow: Nauka; 1975. 188 p, (in Russian)

[10] Ringwood AE. Composition and origin of the moon. In: Hartman WK, Phillips RL, Taylor GJ, editors. Origin of the Moon. Houstan: Lunar Planet. Inst; 1986. pp. 673-698

[11] Schmidt OY. Four Lectures on the Theory of the Origin of the Earth. Publishing House of the USSR Academy of Sciences; 1957. Moscow, 144p

[12] Vityazev AV, Pechernikova GV, Safronov VS. Earth group planets. Origin and early evolution. Moscow: Science. 1990. 296 p

[13] Cameron AGV, Ward W. The Origin of the Moon. Houston: Proceedings of the Seventh Lunar Science Conference; 1976. pp. 120-122

[14] Camp RM. Simulation of the late lunar-forming impact. Icarus. 2004;**168**: 433-456

[15] Taylor SR. The unique lunar composition and its bearing on the origin of the Moon. The Geochim. Et Cosmochim. Acta. 1987;**51**(5):1297-1310

[16] Anfilogov VN, Khachai YV. A possible variant of the differentiation of matter at the initial stage of the formation of the earth. DAN. 2005; **403**(6):803-806

[17] Anfilogov V, Khachay Y. Some Aspects of the Solar System Formation. Briefs of the Earth Sciences. Cham, Heidelberg, New York, Dordrecht, London: Springer; 2015. 75p

[18] Khachay Yu, Hachay O, Antipin A. Dynamics of the earth's-moon system. Geophysics. Okiwelu A, editor. Intechopen. 2018:119-131

[19] Nichols RH Jr. Short lived radionuclides in meteorites: Constraints for space production. Space Science Reviews. 2000;(1-2):113-122

[20] Nyquist LE, Klein T, Shih C-Y, Reese YD. Distribution of short-lived radioisotopes and secondary mineralization. Geochimica et Cosmochimica Acta. 2009;**73**:5115-5136

[21] Safronov VS. Evolution of the Pre-Planetary Cloud and the Formation of the Earth and Planets. Moscow: Nauka; 1969. 244 p

[22] Pechernikova GV, Vityazev AV. Impacts and Evolution of the Early Earth/Catastrophic Impacts of Cosmic bodies. Institute of Geosphere

Dynamics, Russian Academy of Sciences. M.: Academic Book; 2005. pp. 251-265

[23] Tikhonov AN, Lubimova EA, Vlasov VK. On the evolution of the zones of melting in the thermal history of the earth. Dokl. of Akademie of Sciences.USSR. 1969;**188**(2):388-342

[24] Samarsky AA, Moiseenko BD. Computational Mathematics and Mathematical Physics. 1965;**5**(5): 816-827

[25] Stacy FD. Physics of the Earth. Moscow. Mir. 1972. 342 p

[26] Kaula EM. Thermal evolution of the earth and moon growing by planetesimal impacts. Journal of Geophysical Research. 1979;**84**: 999-1008

[27] Shubert G,Turcotte DL, Olson P, Mantle convection in the earth and planets. London: Cambridge University Press 2001; p. 941

Chapter 3

On the Deviation of the Lunar Center of Mass to the East: Two Possible Mechanisms Based on Evolution of the Orbit and Rounding Off the Shape of the Moon

Boris P. Kondratyev

Abstract

It is known that the Moon's center of mass (COM) does not coincide with the geometric center of figure (COF) and the line "COF/COM" is not directed to the center of the Earth, but deviates from it to the South-East. Here, we discuss two mechanisms to explain the deviation of the lunar COM to the East from the mean direction to Earth. The first mechanism considers the secular evolution of the Moon's orbit, using the effect of the preferred orientation of the satellite with synchronous rotation to the second (empty) orbital focus. It is established that only the scenario with an increase in the orbital eccentricity e leads to the required displacement of the lunar COM to the East. It is important that high-precision calculations confirm an increase e in our era. In order to fully explain the shift of the lunar COM to the East, a second mechanism was developed that takes into account the influence of tidal changes in the shape of the Moon at its gradual removal from the Earth. The second mechanism predicts that the elongation of the lunar figure in the early era was significant. As a result, it was found that the Moon could have been formed in the annular zone at a distance of 3–4 radii of the modern Earth.

Keywords: Moon, displacement of center of mass, formation and evolution, gravitation

1. Introduction

At the dawn of modern astronomy, Hevelius and Galileo established that the optical libration of the Moon in longitude leads to a small (5^0–8^0) seeming (for terrestrial observer) oscillations of the figure of our satellite in the East-West direction with a period in the anomalistic month. These oscillations disappear when the Moon is at perigee and apogee. Oscillations of a different kind—optical oscillations in latitude—occur with amplitude $6^0 40'$ and a period of one draconic month with the disappearance of the deviation, when the Moon is at the nodes of the orbit.

If the Moon was absolutely spherically symmetric, these optical librations would not have resulted in additional rotational oscillations of its body. But since due to the interaction with the Earth, the lunar body has tidal bulges, this leads to the appearance of moments of force from external celestial bodies. Newton [1] predicted that deviations of an elongated body of the Moon from the direction to the Earth must lead to real small rotational librations of the satellite relative to the inertial reference system. These small oscillations are called the physical libration of the Moon.

It is necessary to understand that when moving along the orbit of the Moon, its main axis is not directed at the center of mass of the Earth-Moon system and, on the average, at the second (empty) focus of the lunar orbit [2, 3]. The latter will play an important role in our theory.

Due to the proximity of the Moon in our time, the movement of our satellite is studied with such high accuracy that even a small asymmetry of its internal structure must be taken into account. This asymmetry is manifested in that the center of the Moon's mass COM is offset relative to the geometric center of the lunar figure *COF*.

This effect of shift is briefly mentioned in [4, 5]. Using astrometric data, an approximate numerical evaluation of the offset was given in [6] and in a more accurate version in [7]. A new approach based on the analysis of data obtained from the Lunar Laser Ranging experiment allowed in [8] clarifies the parameters of the shift of the Moon's center of mass.

Note that the definition of *COF* depends on the adopted model (sphere, ellipsoid, etc.), so that results of different researchers may be slightly different. However, according to many sources, it is reasonably safe to suggest that two points of the centers on the Moon really do not coincide.

To consider the internal asymmetry of the mass distribution in the lunar body, we introduce a coordinate system with the origin at the center of mass of the Moon, where the *X*-axis is directed (approximately) to the Earth, the *Y*-axis to the left (if viewed from the Earth), and the *Z*-axis—downward. Then, according to the United Lunar Control Network (ULCN), which takes into account the findings of many studies, including information from spacecraft [9], the displacement of the center of the figure relative to the center of mass "COM/COF" is equal to [10]

$$\Delta x \approx -1.71\ km, \quad \Delta y \approx -0.73\ km, \quad \Delta z \approx -0.26\ km. \tag{1}$$

Based on the results of a study of the topography of the lunar surface using laser altimetry from a satellite, the displacement of the "COM/COF" was determined more accurately [11]:

$$\Delta x \approx -1.7752\ km, \quad \Delta y \approx -0.7311\ km, \quad \Delta z \approx -0.2399\ km. \tag{2}$$

As follows from the analysis of observational data (1) or (2), the effect of displacement of the center of the figure relative to the Moon's center of mass includes not only the shift of the center of mass toward the Earth $0.001 \cdot R$ ($R = 1737.10$ km— the average radius of the Moon) but also the spatial deviation of the line "COM/COF" to the North-West. Note that in the literature it often also speaks of the displacement of the center of mass of the Moon relative to the center of its figure; for the observer from the Earth, this shift of the center of mass occurs down (to the South) and to the left (to the East). Then, all the signs in (1) and (2) are reversed. According to (2), the total displacement of the lunar COM is equal to $\Delta \approx 1.935\ km$.

Besides, the shifts (1) and (2) of the center of the Moon's mass are global in nature, and, ultimately, they already include many different factors (see, e.g., [12]). Therefore, in particular, it is impossible to interpret the displacement of the center of mass only as a displacement of the lunar core alone.

Despite the seemingly geometric simplicity of the problem, the offset of the center of the Moon's mass remains an unexplored problem in the lunar science. The importance of this problem is that the Moon is close enough to the Earth and the accuracy of observations of its spin-orbital motion by the method LLR is now so much high that for correct interpretation of these movements it is necessary to take into account many celestial mechanical disturbances, including the indicated internal asymmetry of the Moon's body.

Here, we study the problem of the shift of the Moon's center of mass to the East. To do this, we consider two geometric mechanisms that allow us to explain this important feature of the internal structure of the Moon and shed light on some of the currently controversial features of its evolution and origin (see also [13–15]).

2. Optical libration of the Moon for the observer from the second focus

Instead of the term "the direction of the Moon's surface" often used in references, it is more accurate to speak of the direction of the main lunar axis of inertia, which only in two cases—at the position of the Moon at apogee and perigee—is directed to the center of mass of the Earth-Moon system. To do this, we first consider the optical libration of the Moon in longitude and place the observer in the point of the second (empty) focus of the orbit [2].

Recall that in the first approximation the Moon moves on ellipse (now the eccentricity of the orbit is $e = 0.0549$), and this motion is synchronous, since there is the resonance 1:1 of periods of axial rotation and revolution of the Moon around the Earth. According to the Kepler's first law, the motion is described by the formula

$$r = \frac{p}{1 + e\cos \upsilon}, \quad p = a_1\left(1 - e^2\right). \tag{3}$$

Here, a_1 is the main semiaxis, and e is the eccentricity of an ellipse. The angle of the true anomaly υ is associated with the angle of the eccentric anomaly E

$$\cos \upsilon = \frac{\cos E - e}{1 - e\cos E}. \tag{4}$$

The time that has elapsed since the Moon was at perigee ($E = 0, \upsilon = 0$), until the moment when the angles are equal (E, υ), is equal to

$$t = \frac{(E - e\sin E)}{2\pi} T, \tag{5}$$

where T is the period of revolution on the ellipse. Since the lunar axial angular velocity Ω must be equal the mean motion $n = \frac{2\pi}{T}$, the rotation angle δ of the major axis of inertia of the Moon (see **Figure 1**) in the time t will be

$$\delta = t \cdot \Omega = t \cdot n = \frac{Tn}{2\pi}(E - e\sin E) = E - e\sin E. \tag{6}$$

From the triangle f_1MC (**Figure 1**) follows that

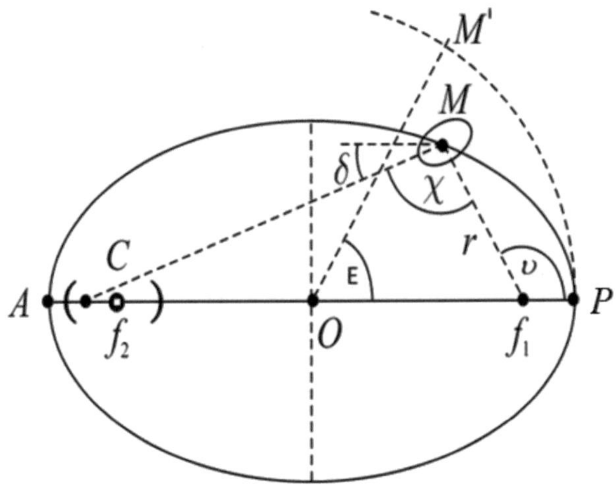

Figure 1.
The large ellipse is the orbit of the Moon M (for clarity, the ellipticity is exaggerated), and P and A are the points of perigee and apogee. The point of active focus f_1 is the center of mass of the Earth-Moon system, and f_2 is the point of the second (passive) focus.

$$\frac{d_{Cf_1}}{\sin\chi} = \frac{r}{\sin(\upsilon - \chi)}, \quad \delta + \chi = \upsilon, \tag{7}$$

so

$$d_{Cf_1} = r \cdot \frac{\sin\chi}{\sin(\upsilon - \chi)} = r\frac{\sin\chi}{\sin\delta}.$$

Then, the distance $\Delta = d_{Cf_2} = 2a_1 e - d_{Cf_1}$ is

$$\frac{\Delta}{a_1} = 2e - \frac{1-e^2}{1+e\cos\upsilon}\frac{\sin\chi}{\sin\delta} = 2e - \frac{1-e^2}{1+e\cos\upsilon}\{\sin\upsilon\ \mathrm{ctg}\,\delta - \cos\upsilon\}. \tag{8}$$

Here, $\mathrm{ctg}\,\delta$ is the function of the angle E (or true anomaly υ)

$$\mathrm{ctg}\,\delta = \mathrm{ctg}\,(E - e\sin E) = \frac{1 + \frac{\sqrt{1-e^2}\sin\upsilon}{e+\cos\upsilon} \cdot \mathrm{tg}\left[\frac{e\sqrt{1-e^2}\sin\upsilon}{1+e\cos\upsilon}\right]}{\frac{\sqrt{1-e^2}\sin\upsilon}{e+\cos\upsilon} - \mathrm{tg}\left[\frac{e\sqrt{1-e^2}\sin\upsilon}{1+e\cos\upsilon}\right]}. \tag{9}$$

Therefore, the required distance $\frac{\Delta}{a_1}$ from the point f_2, which is a continuation of the lunar major inertia axis that crosses the apsidal line, is not, generally speaking, zero and equal to

$$\frac{\Delta}{a_1} = e + \cos E - \mathrm{ctg}\,\delta\sqrt{1-e^2}\sin E \tag{10}$$

Expanding in powers of a small eccentricity gives

$$\frac{\Delta}{a_1} = -\frac{\cos\upsilon}{2}e^2 - \frac{1}{3}\left(1 + \frac{\cos^2\upsilon}{2}\right)e^3 - \frac{\cos\upsilon}{8}(7 - 4\cos^2\upsilon)e^4 + \ldots;\ \Delta \le 0;$$
$$\frac{\Delta}{a_1} = -\frac{\cos E}{2}e^2 + \frac{1}{3}\left(\frac{1}{2} - 2\cos^2 E\right)e^3 - \frac{\cos E}{24}(1 + 8\cos^2 E)e^4 + \ldots;\ \Delta \ge 0. \tag{11}$$

The results of calculations using formula (10) are shown in **Figure 2**.

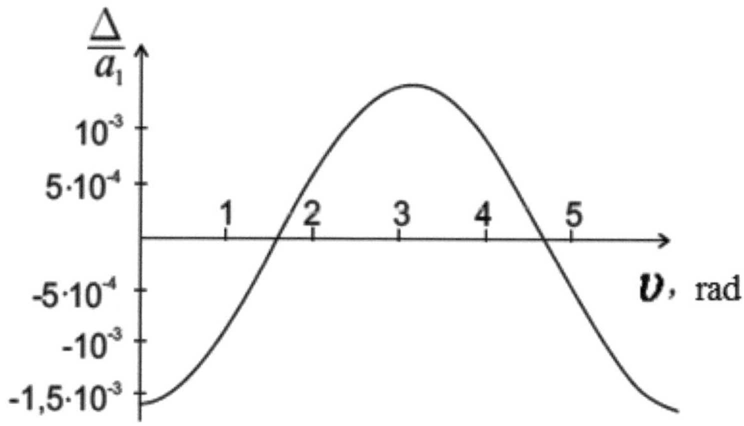

Figure 2.
Graph of deviation $\frac{\Delta}{a_1}$ as a function of the true anomaly υ.

It is important to emphasize that, according to formula (11), the effect of the deviation $\frac{\Delta}{a_1}$ is already in the first approximation proportional to the square of the eccentricity of the Moon's orbit.

Thus, when the Moon is moving on the ellipse around the Earth, the end of the major axis of inertia will be approximately directed to the point of the second focus. Strictly speaking, this end of the axis will perform (without taking into account the very small physical libration of the Moon in longitude) oscillatory motions in the vicinity f_2 in the interval

$$-1.5933 \cdot 10^{-3} \le \frac{\Delta}{a_1} \le 1.4275 \cdot 10^{-3}. \qquad (12)$$

In our era, in a linear measure, this is approximately

$$-612\,km \le \Delta \le 548\,km. \qquad (13)$$

The results of calculations (12) and (13) show a small asymmetry oscillations ($\sim$ 11%) relative to the right and left sides of the point f_2. Emphasize that the physical libration of the Moon in longitude has a very small amplitude and with a large reserve of fits in the interval (13).

3. Resolution alternatives to choose between two options for the lunar orbit evolution

Since Darwin [16], many efforts were made to examine the secular evolution of the Moon's orbit,

but so far it has not been established whether the orbit of the Moon in the past more or less oblate than now. In the literature, this issue is still under discussion. In this regard, the study of the shift of the Moon's center of mass to the East may shed some light on this important issue.

Many researchers agree that gravitational differentiation of the Moon occurred in the early era (see, e.g., [17]), with the result that the Moon's center of mass is slightly ($\sim 0.001 \cdot R$) shifted toward the Earth. We shall not discuss here the question of the gravitational differentiation of the Moon and just to note that one of the reasons for the displacement of the Moon's center of mass to the Earth can be some asymmetry of tidal forces from the Earth into two hemispheres of the Moon

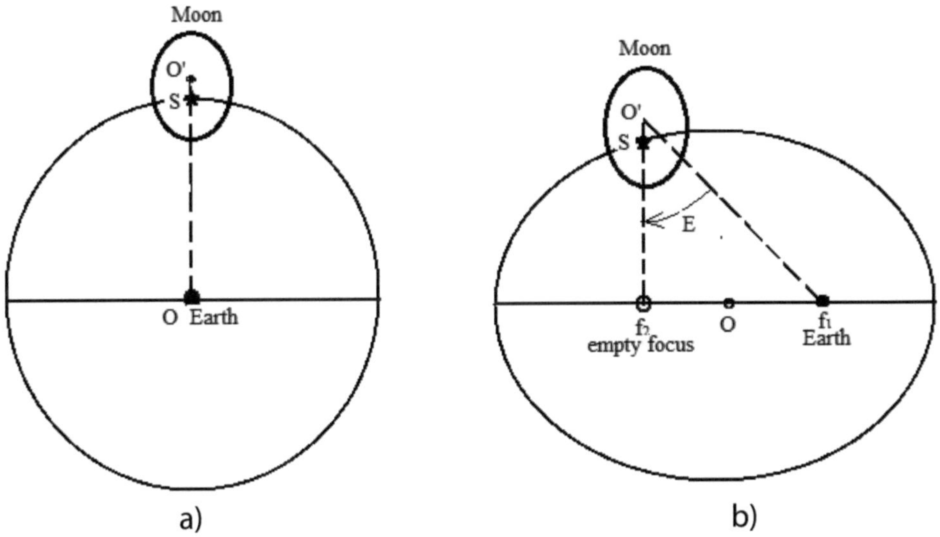

Figure 3.
(a) Orientation of the displaced center of mass S *of the young Moon after the differentiation of the substance of its body. A large circle is the orbit of the Moon in the early epoch, and a small ellipse is the cross section of the Moon. Since the orbit is circular, the focuses* f_1 *and* f_2 *coincide with the center* O. *Relative sizes are not respected. The line* $O'S$ *is directed straight to the Earth; therefore, the Earth's observer would see both points coinciding with each other from Section 2, the motion on the ellipse the line passing through the center of the Moon's figure and its center of mass be directed to the second (empty) focus of the orbit. Therefore, in our era, when the eccentricity of the lunar orbit has increased to its current value* $e = 0.0549$, *we will observe the picture as in b.*
(b) The orientation of the lunar center of mass S *in our era in the first version of the evolution of the Moon's orbit. The large ellipse is the orbit of the Moon, and the small ellipse is the cross section of the Moon. The Earth is in the first focus* f_1 *of the lunar orbit. The angle* E *characterizes the orientation of the Moon COM* S *relative to the direction to Earth.*

(Sect. 4.1). One of the manifestations of the offset center of mass can be a different thickness of crust in the near side and the far side of the Moon [18].

Thus, the core of the Moon was formed during the gravitational differentiation, and then under the influence of a small asymmetry of tidal forces, the process of displacement of the lunar center of mass toward the Earth began to occur. This offset COM for the Earth observer can be characterized by the orientation angle E between the line $''COF/COM''$ and the direction to the center of the Earth (**Figure 3b**).

3.1 On the difference on tidal forces from the Earth in near and far lunar hemispheres

Assuming that the differentiation of the Moon occurred (according to cosmogonic times) rather quickly, it is necessary to require that the shift of the lunar center of mass toward the Earth occurred even before the Moon hardened.

The real cause of the displacement of the Moon's center of mass to the Earth could be some asymmetry of tidal forces. Let us perform the required calculations. After the capture of the Moon in resonance 1:1, it was possible to talk about near and far of its hemispheres. It is clear that the forces in the nearest and farthest points are, respectively, equal to

$$F_1 \approx \frac{2GM_\oplus}{R_0^2} x\left(1+\frac{3}{2}x\right),$$
$$F_2 \approx \frac{2GM_\oplus}{R_0^2} x\left(1-\frac{3}{2}x\right), \quad x = \frac{R}{R_0}, \tag{14}$$

where R_0 is the distance between the centers of the Earth-Moon and R is the distance from the center of the Moon to the near (far) points of its surface. The difference of these forces will be

$$\Delta F_{\oplus} = F_1 - F_2 \approx \frac{6GM_{\oplus}}{R_0^2} x^2. \tag{15}$$

In the era of its formation, the Moon could be much closer to Earth than in our era (see, e.g., [16, 18, 19]). Due to the proximity to the Earth of the young Moon, the difference in tidal forces (15) in both lunar hemispheres was much more in the early era than it is now. In the era of the differentiation of the Moon, it was this difference in tidal forces (15) that caused the displacement of the center of mass of the Moon toward the Earth. Based on these provisions, we note that the very solution to the question of the displacement of the Moon's *COM* to the East is closely related to the further secular evolution of its form and orbit. In particular, to find out how the lunar *COM* would be located relative to the Earth's observer in the modern era, when its orbit evolved and eccentricity acquired modern significance, consider two possible options with the initial eccentricity of the young Moon orbit.

3.2 The first version: the evolution of the lunar orbit with increase in its eccentricity

First, suppose that in the early epoch the orbit of the Moon was more circular than in our epoch. Consequently, during the secular evolution, the Moon's orbit became more and more eccentric, up to its modern value of eccentricity $e = 0.0549$.

Recall now that the Moon's COM, already shifted toward the Earth, after the solidification of the lunar body will be fixed relative to its main axes of inertia. Since in the early epoch the orbit of the Moon was almost circular, the line connecting the geometrical center of the figure of the Moon and its center of mass was directed exactly to the Earth (**Figure 3a**).

However, since in this version of the secular evolution the orbit of the Moon becomes more eccentric, two foci appear (**Figure 3b**). In accordance with the laws of celestial mechanics, as we know

From **Figure 3b**, it can be seen that, for the observer from the Earth (point f_1), the center of mass S will now be located on the left (to the East) from the direction to the center of the Moon (see also **Figure 5**). *Thus, in the first variant of the evolution of the Moon's orbit, the modern Earth's observer, in accordance with* **Figure 3b**, *will see the Moon's center of mass displaced to the left (to the East) from the direction to the center of the figure. It is this location of the center of mass of the Moon relative to the center of its figure that we observe in our era.*

The contribution of this mechanism to the displacement of the Moon's center of mass to the East will be made in Section 4.

3.3 The second version of the evolution: from more flattened to less flattened lunar orbit

If we assume that the orbit of the young Moon was more eccentric in the early era than it is now, that is, during the secular evolution, the Moon's orbit was rounded; then in our era, when the orbital eccentricity decreased to the current value $e = 0.0549$, instead of **Figure 3b**, we will see the location of the center of mass of the Moon, as shown in **Figure 4**.

Thus, **Figure 4** shows that in the second version of the evolution of the orbit a modern observer from the Earth would see that the center of mass of the Moon is

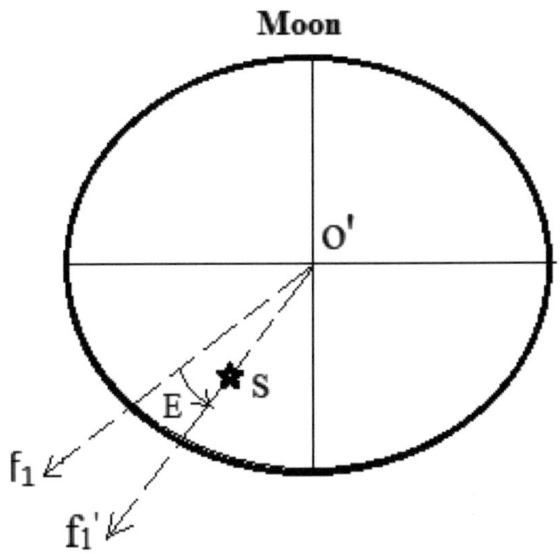

Figure 4.
The final configuration of the location of the lunar COMS in the second version of the evolution of its orbit. The line $O'f_1$ sets the direction (for the Earth observer) to the center of mass of the Moon in the early era, and the line $O'f'_1$ sets the direction to the center of the figure of the Moon for the observer from the Earth in our time. In this version, the observer would see that the center of mass of the Moon S is shifted to the right (to the West, as indicated by the arrow) from the average direction to the center O′ of the figure of the Moon.

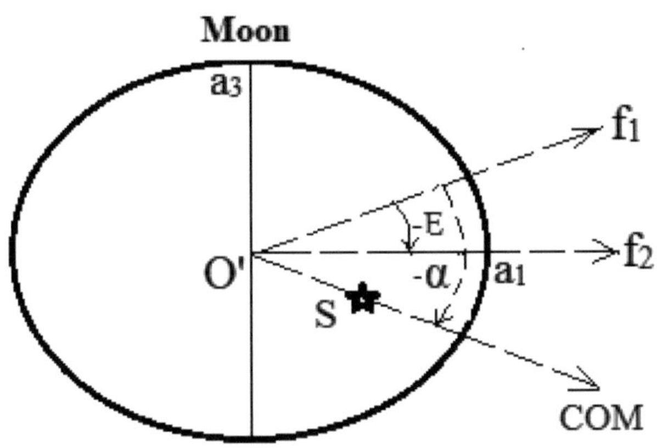

Figure 5.
Elongated figure of the Moon in the early era (its cross section of the plane $x_2 = 0$ is shown by the ellipse with the semiaxes $a_1 > a_3$). The arrows represent the directions from the center O' of the Moon to both foci f_1 and f_2 of its orbit around the Earth (the Earth in focus f_1), as well as to the center of the mass S of the Moon. The angle α between the line on S and the line to f_1 measures orientation of the Moon's center of mass, and the angle E between the directions to the first focus f_1 and second focus f_2 measures the deflection of an ellipse from a circle.

shifted to the right (to the West) from the direction to the center of the figure. *However, this is contrary to observations, so the second version of the evolution must be discarded.*

4. Correction factor to mechanism of orbit evolution

Let us consider again (**Figure 1**) the motion of a satellite in an elliptical orbit around a body of greater mass. The equation of an ellipse is given by formula (3). From the triangle $O'f_1f_2$ by the sine theorem, we find the relation

$$\sin E = \frac{2e \sin \upsilon (1 + e \cos \upsilon)}{1 + e^2 + 2e \cos \upsilon}. \tag{16}$$

Then, the average angle E is given by the integral

$$\langle E \rangle = \frac{1}{\pi} \int_0^{\pi} \arcsin \left[\frac{2e \sin \upsilon (1 + e \cos \upsilon)}{1 + e^2 + 2e \cos \upsilon} \right] \tag{17}$$

In particular, for the Moon's orbit, the current value of eccentricity is equal $e \approx 0.0549$, and formula (17) gives

$$\langle E \rangle \approx 0.0700. \tag{18}$$

Taking into account (18), in the framework of the first variant of the evolution mechanism of the lunar orbit from the circle to the ellipse with the modern value of eccentricity, we find that the ratio of the average angle $\langle E \rangle$ to the angle $\mathrm{arctg} \frac{\Delta y}{\Delta x}$ will be

$$\kappa = \frac{\langle E \rangle}{\mathrm{arctg}(0.7311/1.7752)} \approx 0.18. \tag{19}$$

Therefore, the first orbital evolution mechanism helps to explain approximately 18% of the observed current Moon's offset *COM* to the East. In the linear measure, it is

$$|\Delta y| \approx 0.132 \ km. \tag{20}$$

We emphasize that the conclusion of the theory that evolution of the orbit of the Moon occurred with increasing eccentricity is consistent with the fact that at the present time the eccentricity of the orbit of the Moon is really growing and, therefore, in the past it was less than today [20, 21] (see also [22–25]).

Besides, the following should be noted. As is well known, due to perturbations, all elements of the lunar orbit are subject to periodic perturbations [20, 26]. Thus, for several thousand years, the eccentricity of the Moon's orbit changes due to solar perturbations in the range from 0.0255 to 0.0775. However, here we do not consider the periodic perturbations: throughout in this chapter, we are talking about tidal secular change in the average eccentricity of the Moon's orbit, which is now equal $e \approx 0.0549$.

5. Second mechanism of displacement of the Moon's center of mass to the East

Because of proximity of the Moon to Earth during an early era, which is offered by many researchers, the main factor of formation for the Moon is a tidal force from our planet. In the tidal field of the Earth, the figure of the early Moon stretched out, which was also facilitated by its capture in spin-orbit resonance 1:1. Therefore, for our approximate calculations, we can simulate the figure of the Moon using the elongated (toward the Earth) spheroid with the semiaxes $a_1 > a_2 = a_3$. The equation of the surface of this spheroid in Cartesian coordinates $Ox_1x_2x_3$ is

$$\frac{x_1^2}{a_1^2} + \frac{x_2^2 + x_3^2}{a_3^2} = 1. \tag{21}$$

The main symmetry semiaxis a_1 of this spheroid was initially directed exactly to the Earth.

Let us consider **Figure 5**. Due to the small orbit eccentricity, the angle E between the main axis of the Moon's figure and the direction to f_1 was also initially small. However, in the evolution of the Moon's orbit from the less eccentric to the more eccentric, as was shown in the first mechanism, the angle E will increase monotonically. This factor changes the orientation of the figure of the Moon relative to the observer on the Earth, and the angle α will also increase. From a geometrical point of view, during the evolution of the lunar orbit, the angle E can change only in the interval of values $0 \leq E \leq 2e \approx 0.11$. Moreover, taking into account the averaging performed above (see form. (18)), the right part of the interval will be adjusted

$$0 \leq E \leq 0.070. \tag{22}$$

In addition, although the angle α can vary from zero (in the early era of lunar evolution) up to the current value $\alpha_0 = \arctan\left(\frac{0.7311}{1.7752}\right) \approx 0.39$, but also taking into account the action of the first mechanism, the interval will be changed:

$$0 \leq \alpha \leq \alpha_0, \tag{23}$$

where

$$\alpha_0 = \arctan\left(\frac{0.7311 - 0.1243}{1.7752}\right) \approx 0.329. \tag{24}$$

We emphasize that because of inequalities (22) and (23), the center of mass of the Moon will have that arrangement which is shown in **Figures 3b** and 5.

The problem consists in studying dependence between the angle α and the changing form of the Moon during the secular evolution in the gravitational tidal field of the Earth.

6. Differential equation for evolution of the angle α

As you know (see, e.g., [27]), a change in the shape of an ellipsoidal body can be described by a linear velocity field. In particular, the evolution of the prolate spheroid (21) in the moving frame of reference, whose axes coincide with the main axes of this body at any time, can be represented by the velocity field:

$$u_1 = \frac{\dot{a}_1}{a_1} x_1, \quad u_2 = \frac{\dot{a}_2}{a_2} x_2, \quad u_3 = \frac{\dot{a}_3}{a_3} x_3. \tag{25}$$

Here, the point above denotes the time derivative $\frac{d}{dt}$. Since for incompressible figures the condition of volume preservation should be fulfilled (in this case—for the volume of the prolate spheroid (21)), we have the additional ratio

$$\text{div } \mathbf{u} = \frac{\dot{a}_1}{a_1} + 2\frac{\dot{a}_3}{a_3} = 0. \tag{26}$$

In the velocity field (25), the Moon's shape will always remain a second-order surface, and the streamlines will be represented by pieces of hyperboles (**Figure 6**).

Owing to symmetry, the elongation of the spheroid (21) is described by the only polar oblateness $\varepsilon = 1 - \frac{a_3}{a_1}$. Consider changing ε for the Moon's shape. In this case

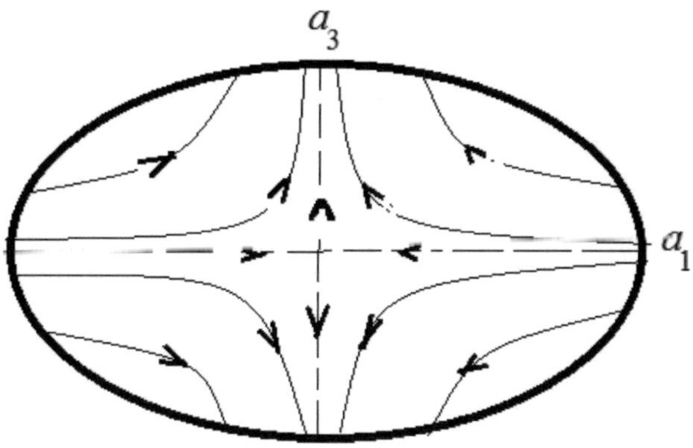

Figure 6.
Streamlines at deformation of the Moon's shape (the section is shown by ellipse). Arrows depict the direction of deformation at the stage of rounding the figure in the early era of evolution.

two components (first and third) of the velocity field in (25) taking into account a condition of incompressibility (26) will take the form

$$u_1 = \gamma x_1, \quad u_3 = -\frac{1}{2}\gamma x_3; \quad \gamma = \frac{1}{a_1}\frac{da_1}{dt}. \tag{27}$$

In the plane Ox_1x_3, the condition $x_2 = 0$ is satisfied, and expressions for angles E and α, (see **Figure 5**) will be equal:

$$E = -\mathrm{arctg}\frac{x_3}{x_1}, \quad E - \alpha = -\mathrm{arctg}\frac{x_3'}{x_1'}. \tag{28}$$

Here, (x_1, x_3) and (x_1', x_3') are the coordinates of the points of intersection of the Moon's surface by the rays $O'f_1$ and $O'S$, respectively. Therefore,

$$\alpha = \mathrm{arctg}\frac{x_3'}{x_1'} - \mathrm{arctg}\frac{x_3}{x_1}. \tag{29}$$

Differentiating expression (29) with respect to time t, we find

$$\dot{\alpha} = \frac{x_1'\dot{x}_3' - x_3'\dot{x}_1'}{x_1'^2 + x_3'^2} - \frac{x_1\dot{x}_3 - x_3\dot{x}_1}{x_1^2 + x_3^2}; \tag{30}$$

By substituting in (30) the components of the velocity field (27), we obtain

$$\dot{E} = -\frac{3}{4}\gamma \sin 2E; \quad \dot{E} - \dot{\alpha} = -\frac{3}{4}\gamma \sin 2(E - \alpha). \tag{31}$$

Thus, the derivative of the angle α will be equal to

$$\dot{\alpha} = -\frac{3}{4}\gamma[\sin 2E - \sin 2(E - \alpha)]. \tag{32}$$

More convenient than (32), below will be the next form of differential equation:

$$\frac{d\alpha}{dt} = -\frac{3}{2}\gamma \sin\alpha \cdot \cos(2E - \alpha). \tag{33}$$

7. Solution of Eq. (33)

Let us turn to the analysis of the differential equation (33) and transform the derivative $\frac{d\alpha}{dt}$:

$$\frac{d\alpha}{dt} = \frac{d\alpha}{d\varepsilon}\frac{d\varepsilon}{dt}. \tag{34}$$

As

$$\frac{d\varepsilon}{dt} = \frac{d}{dt}\left(1 - \frac{a_3}{a_1}\right) = \frac{a_3\dot{a}_1 - a_1\dot{a}_3}{a_1^2}, \tag{35}$$

therefore, in agreement with (34),

$$\frac{d\varepsilon}{dt} = \frac{3}{2}\gamma(1-\varepsilon). \tag{36}$$

Substituting (36) in (34) and then the result in (33), we have

$$\dot{\alpha} = \frac{3}{2}\gamma(1-\varepsilon)\frac{d\alpha}{d\varepsilon} = -\frac{3}{2}\gamma\sin\alpha\cos(2E-\alpha). \tag{37}$$

As a result, the differential equation for the angle α takes the form

$$\frac{d\alpha}{d\varepsilon} = -\frac{\sin\alpha\cos(2E-\alpha)}{1-\varepsilon}. \tag{38}$$

Separating the variables in (38) and integrating and taking into account the auxiliary formula

$$\int\frac{d\alpha}{\sin\alpha\cos(2E-\alpha)} = \frac{1}{\cos 2E}\ln\frac{\sin\alpha}{\cos(2E-\alpha)}, \tag{39}$$

we obtain a solution for equation (38) in the form

$$\frac{1}{\cos 2E}\ln\frac{\sin\alpha}{\cos(2E-\alpha)} = C + \ln(1-\varepsilon), \tag{40}$$

where C is the integration constant. Potentiating expression (40), we find the solution in the form

$$\varepsilon(\alpha, E) = 1 - C\cdot\exp\left\{\left[\frac{\sin\alpha}{\cos(2E-\alpha)}\right]^{\frac{1}{\cos 2E}}\right\}. \tag{41}$$

8. Analysis of the solution (41) and estimation of the elongation of the lunar figure in early era

In formula (41), the constant integration C is defined by the known observational data. As in the modern epoch of tidal evolution of the Moon the supplemented relations

$$\varepsilon \approx 0.0125, \quad E = 0.07, \qquad \alpha = \arctan\frac{0.7311 - 0.1243}{1.7752} \approx 0.32937, \tag{42}$$

then the formula (41) gives

$$C \approx 0.713. \tag{43}$$

Thus, the solution of equation (41) will get in the form

$$\varepsilon(\alpha, E) = 1 - 0.713 \cdot \exp\left(\left[\frac{\sin\alpha}{\cos(2E - \alpha)}\right]^{\frac{1}{\cos 2E}}\right). \tag{44}$$

Formula (44) represents the solution of the problem: it describes the change in the Moon's oblateness ε during the tidal evolution and establishes the dependence between ε and the angle α. Recall that α is the angle between the directions (from the center of the Moon) to the first focus of the orbit and the Moon's *COM*. As we already know, in the course of evolution, the angle α varied (in radians) within the limits given in (23).

The graphic image of the function of two variables from (44) is shown in **Figure 7**.

Graphs for the two extreme values of the angle E are shown in **Figure 8**. As seen in **Figure 8**, the oblateness ε of the figure is very little depending on the angle E. Moreover, in the initial era, ε for all E has the same value and could not exceed the value

$$\varepsilon = 1 - \frac{a_3}{a_1} \approx 0.285. \tag{45}$$

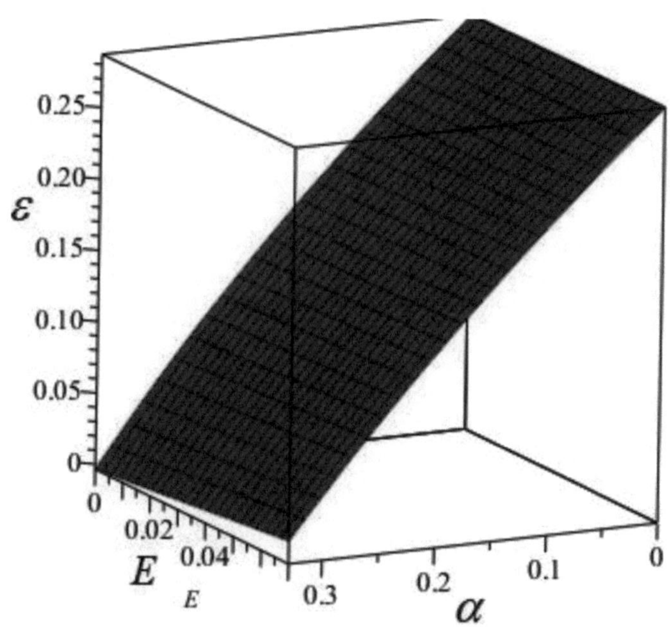

Figure 7.
3D image of the function $\varepsilon(\alpha, E)$. The angle α is set in radians. The oblateness of shape of the Moon $\varepsilon(\alpha, E)$ in the early era very little depends on the value of the parameter E, and its value does not exceed $\varepsilon \approx 0.285$.

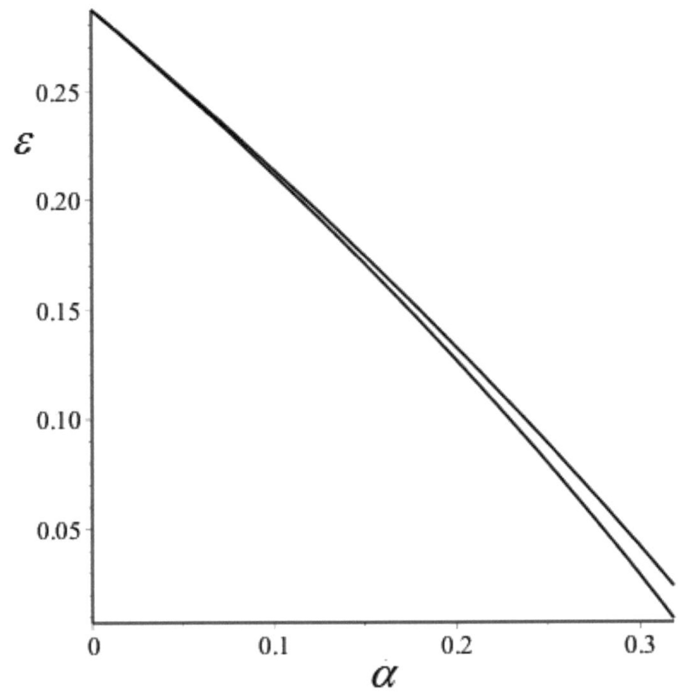

Figure 8.
The dependence of the oblateness ε of the Moon shape from the angle α between the line "COM/COF" of the Moon and the mean direction to the Earth. The graph shows the change ε during the tidal evolution. Two extreme angle values E = 0.0 (upper curve) and E = 0.07 are taken for comparison. The beginning and the end of the evolutionary process correspond to the values α ≈0 and α ≈0.329.

Thus, the second mechanism explains both the displacements of the center of mass of the Moon to the East and predicts that the oblateness of the Moon in the early era could not exceed the value $\varepsilon \approx 0.285$.

9. Some consequences: how close to the earth could the Moon be formed

Above we established that on the known shift of the Moon's center of mass to the East, we can find the oblateness (45), which the Moon could have in the epoch of its formation. The corresponding spheroid eccentricity will be equal to

$$e \approx 0.70. \tag{46}$$

Proceeding from (46) and using the theory of tidal equilibrium figures, it is possible to estimate how close to each other might be the Earth and the Moon in the early era. For this purpose, without loss of generality, we assume that the satellite is uniform (at the Moon, as we know, and now concentration of substance very small), and its mass in comparison with the mass of the Earth can be neglected. Then, in the tidal approach for the potential of the Earth, the equation of hydrostatic equilibrium of the satellite with synchronous rotation has the first integral [28]:

$$\frac{p}{\rho} + const = \varphi + \frac{1}{2}\Omega^2\left(3x_1^2 - x_3^2\right); \qquad \Omega^2 = \frac{GM_\oplus}{R_\oplus^3}. \tag{47}$$

Here, p is the pressure, ρ is the density, φ is the quadratic internal gravitational potential of the satellite, Ω is the angular velocity rotation of the satellite, and $R_\oplus$ is the distance between the centers of the Earth and the Moon. For satellite with the form of the prolate spheroid (21), we have [27]

$$\begin{aligned} \varphi &= \pi G\rho\left[I - A_1x_1^2 - A_3\left(x_2^2 + x_3^2\right)\right]; \\ A_1 &= \frac{1-e^2}{e^3}\ln\frac{1+e}{1-e} - 2\frac{1-e^2}{e^2}; \\ A_3 &= \frac{1}{e^2} - \frac{1-e^2}{2e^3}\ln\frac{1+e}{1-e}. \end{aligned} \tag{48}$$

The internal pressure of the equilibrium figure should also be a quadratic function from the coordinates

$$p = p_0\left(1 - \frac{x_1^2}{a_1^2} - \frac{x_2^2 + x_3^2}{a_3^2}\right). \tag{49}$$

From the first integral (47) is possible to find a square of angular velocity rotation of satellite

$$\frac{\Omega^2}{\pi G\rho} = 2\frac{A_1 - (1-e^2)A_3}{4-e^2}. \tag{50}$$

Since

$$\frac{\Omega^2}{2\pi G\rho} = \frac{M_\oplus}{2\pi\rho R_\oplus^3} = \frac{\kappa}{x^3}, \tag{51}$$

where we have identified the following characters

$$x = \frac{R_\oplus}{R_\oplus}; \quad \kappa = \frac{2}{3}\frac{\rho_\oplus}{\rho} \approx 1.09875, \tag{52}$$

the ratio (51) can be represented as

$$\frac{\kappa}{x^3} = \frac{A_1 - (1-e^2)A_3}{4-e^2}. \tag{53}$$

Substituting the value e from (46) into the right-hand side (53), we obtain the cubic equation

$$\frac{\kappa}{x^3} = 0.0324, \tag{54}$$

from which we find the required distance

$$x \approx 3.24. \tag{55}$$

Thus, the Moon with oblateness (45) could form at a very close distance from the Earth: at a distance of only three and a quarter of the mean radii of the modern Earth. This result slightly corrects the one we received earlier [15].

Note that the prolate spheroid with meridional eccentricity (45) is a stable figure of equilibrium. In fact, the instability of this type of figure occurs only when $e \geq 0.883$ (see, e.g., [28]).

10. Discussion and conclusions

Here, it is necessary to add the following. As is well known, in the problem of secular perturbations, the perturbation function is replaced by its secular part. The influence of the Sun leads only to periodic perturbations of the eccentricity of the lunar orbit, which we do not take into account here. In this chapter, we ignore periodic oscillations and consider only tidal secular changes in the average eccentricity of the lunar orbit.

As for the tidal influence of the Sun on the figure of the Moon, it turns out to be insignificant compared to the influence of the Earth. Indeed, the ratio of force $\Delta F_{\odot}$ to force $\Delta F_{\oplus}$ from (15) is equal to

$$\frac{\Delta F_{\odot}}{\Delta F_{\oplus}} = \frac{M_{\odot}}{M_{\oplus}} \left(\frac{R_{\oplus}}{R_{\odot}}\right)^4 \approx 10^{-5}.$$

Therefore, to solve the posed problem within the framework of our model, the influence of the Sun can be neglected.

In the theory of the tidal evolution of the Moon's orbit and its form, we encounter problems that are difficult to give exact answers. Above, we examined some of the conclusions from those observational facts that the center of mass of the Moon is slightly shifted to the East. Two geometrical mechanisms have been developed to explain this shift.

The first mechanism considers the secular evolution of the Moon's orbit, using the effect of the preferred orientation of the satellite with synchronous rotation to the second orbital focus. According to this mechanism, *only the scenario of secular evolution of the orbit with the increase of eccentricity leads to the desired offset of the center of the Moon's mass to the East*. It is important to note that this conclusion that the evolution of the Moon's orbit occurred with an increase e is consistent with the fact that at present the eccentricity of the lunar orbit is indeed increasingly, and therefore in the past, it was less than today [20, 21] (see also [22–25]).

To fully explain the displacement of the center of the Moon's mass to the East, a second mechanism was developed, which takes into account the influence of tidal changes in the shape of the Moon as it gradually moves away from the Earth. The essence of the second mechanism is fully consistent with the fact that the distance between Earth and Moon is now really increasing and the Earth's spin is slowing in reaction.

In addition, the second mechanism predicts that the Moon's figure flattening in the early era was very significant and reached the value of $\varepsilon \approx 0.285$. In turn, based on the theory of tidal equilibrium figures, it allowed us to estimate how close to Earth could the Moon be formed as an astronomical body. According to formula (55), the Moon was formed in the ring zone at a distance of 3–4 medium radii of the present Earth. This result seems to be consistent with the modern view that the Moon was formed as a result of a gigantic impact in the immediate vicinity of the proto-Earth.

Since the formation of the Moon as a celestial body and so far the Earth-Moon system has been and remains a binary planet, the physical laws of its development have always been the same. In the early era, however, the tidal forces between the Earth and the Moon were much more important. Indeed, now the tidal force has very little effect on the Moon, because of which it is removed from the Earth for only 3.8 cm per year. However, studying the evolution of the moon still requires a great effort of researchers.

In summary, we can say that the method presented here really allows to take into account additional observational facts in the structure of the Moon. We have shown that from the hidden fact that in our era there is a slightly shift of the center of the Moon's mass to the East, and not to the West, you can get valuable information about the evolution of the orbit of the Moon and its shape. This finding supports the scenario [29] that the Moon could be formed about 4.5 billion in the surrounding "donut" from the hot gas that appeared after the collision of Theia with proto-Earth.

Author details

Boris P. Kondratyev[1,2,3]

1 Sternberg Astronomical Institute, M.V. Lomonosov Moscow State University, Russia

2 Faculty of Physics of the M.V. Lomonosov Moscow State University, Russia

3 Central Astronomical Observatory at Pulkovo, Russia

*Address all correspondence to: work@boris-kondratyev.ru

References

[1] Newton I. Mathematical principles of natural philosophy. In: Bernard Cohen I, Whitman A, editors. A Guide to Newton's Principia. University of California Press; 1999

[2] Kondratyev BP. On one inaccuracy of Isaac Newton. Kvant. 2009;**5**:38

[3] Murray K, Dermott S. Dynamics of the Solar System. M.: FIZMATLIT, 2010; p. 588 (trans. from English).

[4] Yakovkin AA. The radius and shape of the moon. Bulletin AOE. 1934;**4**

[5] Bohme S. Bearbeitung der Aufnahmen von F. Hayn zur Ortsbestimmung des Mondes. Astronomische Nachrichten. 1953;**256**: 356

[6] Shakirov KS. The influence of the internal structure of the moon on its rotation. Izvestia AOE. 1963;**34**

[7] Lipsky YN, Nikonov VA. The position of the center of the figure of the moon. Astronomicheskii Zhurnal. 1971; **48**:445

[8] Calame O. Free librations of the Moon determined by an analysis of laser range measurements. Moon. 1976;**15**:343

[9] Archinal BA, Rosiek MR, Kirk RL, Redding BL. Completion of the Unified Lunar Control Network 2005 and Topographic Model. Virginia: US Geological Survey. 2006

[10] Iz H, Ding XL, Dai CL, Shum CK. Polyaxial figures of the Moon. Journal of Geodesy. 2011;**1**(4):348

[11] Barker MK, Mazarico E, Neumann GA, Zuber MT, Kharuyama J, Smith DE. A new lunar digital elevation model from the Lunar Orbiter Laser Altimeter and SELENE Terrain Camera. Icarus. 2016;**273**:346

[12] Lemoine FG, Goossens S, Sabaka TJ, Nicholas JB, Mazarico E, Rowlands DD, et al. GRGM900C: A degree-900 lunar gravity model from GRAIL primary and extended mission data. GeoRL. 2014;**41**: 3382. DOI: 10.1002/2014GL060027

[13] Kondratyev BP. The deviation of the lunar center of mass to the east of the direction toward the earth. A Mechanism Based on Orbital Evolution. Astronomy Reports. 2018;**62**(8):542. DOI: 10.1134/S106377291808005X

[14] Kondratyev BP. The deviation of the lunar center of mass to the east of the direction toward the earth. A Mechanism Based on Rounding of the Figure of the Moon. Astronomy Reports. 2018; **62**(10):705. DOI: 10.1134/S106377 2918100062

[15] Kondratyev BP. On the deviation of the lunar center of mass to the East. Two possible mechanisms based on evolution of the orbit and rounding off the shape of the Moon. Astrophysics and Space Science. 2018;**186**(186)

[16] Darwin GH. Tidal Friction in Cosmogony, Scientific Papers 2. Cambridge University Press; 1908

[17] Urey HC. Chemical evidence relative to the origin of the solar system. MNRAS. 1966;**131**:212

[18] Wieczorek MA, Neumann GA, Nimmo F, Kiefer WS, Taylor GJ, Melosh HJ, et al. The crust of the Moon as seen by GRAIL. Science. 2013;**339**(6120):671. DOI: 10.1126/science.1231530

[19] Zhong S. Origin and Evolution of the Moon. 2014. 2014IAUS, 298, 457Z. DOI: 10.1017/S1743921313007229

[20] Macdonald GJF. Tidal friction. Reviews of Geophysics. 1964;**2**:467. DOI: 10.1029/RG002i003p00467

[21] Goldreich P. History of the lunar orbit. Reviews of Geophysics and Space Physics. 1966;**4**:411

[22] Simon JL, Bretagnon P, Chapront J, Chapront-Touze M, Francou G, Laskar J. Numerical expressions for precession formulae and mean elements for the moon and planets. A&A. 1994; **282**:663

[23] Chapront J, Chapront-Touzé M, Francou G. A new determination of lunar orbital parameters, precession constant and tidal acceleration from LLR measurements. A&A. 2002;**387**:700

[24] Laskar J, Fienga A, Gastineau M, Manche H. La2010: A new orbital solution for the long-termmotion of the Earth. A&A. 2011;**532**(**A89**):15

[25] Folkner WM, Williams JG, Boggs DH, Park RS, Kuchynka P. The Planetary and Lunar Ephemerides DE430 and DE431. The Interplanetary Network Progress Report. 42-196. 2014. pp. 1-81

[26] Deprit A. The movement of the moon in space. In: Kopal Z, editor. Physics and Astronomy of the Moon. New York and London: Academic Press; 1971

[27] Kondratyev BP. Dinamika ellipsoidal'nykh gravitiruiushchikh figure. Moscow: Nauka; 1989

[28] Chandrasekhar S. Ellipsoidal Equilibrium Figures. New Haven and London: Yale University Press; 1969

[29] Simon SJ, Stewart ST, Petaev MI, Leinhardt ZM, Mace MT, Jacobsen SB. The Origin of the Moon Within a Terrestrial Synestia. JGR. 2018. DOI: 10.1002/2017JE005333

Chapter 4

New Principles of Monitoring Seismological and Deformation Processes Occurring in the Moon Rock Massive

Olga Hachay and Oleg Khachay

Abstract

Currently, the interest in studying the processes occurring in other planets surrounding the Earth is becoming increasingly important. The Moon-satellite planet is the closest to the planet Earth, and therefore, it makes sense to organize a system for studying it first and foremost, incorporating the most advanced ideas about the physics of processes in rock massive, which are also used in terrestrial conditions. In this paper, new ideas on the organization of seismological and deformation monitoring are set out, based on the results obtained for the rock massive of the Earth and the theoretical ideas presented in the works of I. Prigogine and S. Hawking.

Keywords: Moon rock massive, seismological and deformation processes, new principles of monitoring

1. Introduction

In recent decades, a new science was born—the physics of non-equilibrium processes associated with such concepts as irreversibility, self-organization, and dissipative structures [1]. It is known that irreversibility leads to many new phenomena, such as the formation of vortices, vibration chemical reactions, or laser radiation. Irreversibility plays a significant constructive role. It is impossible to imagine life in a world devoid of interrelations created by irreversible processes. The prototype of the universal law of nature is Newton's law, which can be briefly formulated as follows: acceleration is proportional to force. This law has two fundamental features. It is deterministic: since the initial conditions are known, we can predict movement. And, it is reversible in time: there is no difference between predicting the future and restoring the past; movement to a future state and reverse movement from the current state to the initial state are equivalent. Newton's law is the basis of classical mechanics, the science of the motion of matter, of trajectories. Since the beginning of the twentieth century, the boundaries of physics have expanded significantly. Now, we have quantum mechanics and the theory of relativity. But, as we shall see from the sequel, the basic characteristics of Newton's law—determinism and reversibility in time—are preserved. Is it possible to modify the very concept of physical laws so as to include in our fundamental description of the nature of irreversibility, events and the arrow of time? The adoption of such a program entails a thorough revision of our formulation of the laws of nature, and it became possible due to the remarkable successes associated with the ideas of instability and chaos [1, 2].

Returning to the results obtained for unstable mountain terrestrial massive, we can note that monitoring studies should be conducted in an active mode, i.e., there should be a source of excitation (seismic or of other nature), and the response of the rock massive is recorded for a not very long time; then, the effect should be repeated, and for this process, as a result, phase diagrams of the rock massive can be constructed.

2. The structure, composition, state, and some methods of their research and their comparison of the Earth and the Moon

This brief review is based on the book [3]. The Moon, devoid of atmosphere and hydrosphere, is a unique repository of traces of the cosmic history of the planets, the key by which people hoped to penetrate into geological secrets. Long-term ground studies of the Moon using astronomical and astrophysical methods have allowed us to obtain numerous data on the laws of its movement and physical conditions on the lunar surface, but this was not enough. Only the development of space technology made it possible to connect the traditional Earth sciences—geology, geochemistry, geophysics, and geodynamics—with powerful methods of observation and data analysis to the studies of the Moon. Successful flights to the Moon and its photographing at close range allowed, first of all, strengthening the geological exploration of the Moon by analyzing its photographs. The delivery of samples of lunar matter opened up unprecedented possibilities for studying the age and composition of the rocks of the Moon using geochemistry methods and allowed to establish the sequence of events in the evolution of the Moon. The first soft landing on the Moon, carried out in 1966 by the Soviet automatic station Luna-9, opened a period of active research directly on the surface of the Moon. The Soviet automatic stations were the first to perform a soft landing in the sea and continental areas of the Moon, photographing the surface, measuring the density and strength of lunar soil, drilling the surface layer, and measuring the physical properties of the soil along long profiles.

Measurements of mechanical as well as some other physical characteristics of the soil at various points on the lunar surface were carried out by American automatic stations "Surveyor." A large complex of observations on the surface, the selection and delivery of lunar samples, and the measurement of the physical fields on the Moon were made by the American Apollo space missions. The book [3] summarizes the history, methods, and results of complex physical studies of the lunar soil, the crust, and the deep depths of the moon, carried out in the past 40 years. An analysis of all the accumulated data made it possible to assess the state of the substance of the Earth's interior, its composition, and evolution features and compare the properties of the Earth and the Moon. The mining of deeply hidden underground minerals, the prediction of earthquakes and volcanic eruptions, and the preservation of the ecological balance in nature are currently important tasks facing humanity. For this, it was necessary to understand the Earth and to reveal the secrets of its history and device. Geophysics helps to penetrate into its supercompressed and red hot depths. However, not all problems can be solved while on Earth.

We practically do not analyze here data concerning the composition and age of the rocks of the Moon; this important and extensive topic requires special consideration. Conversely, the entire sections were allocated to describe lunar soil studies, which were carried out according to a special technique, in large volumes, and are of great scientific interest both in terms of improving space technology and for extrapolating open patterns to great depths [5–10].

The most important information about the structure and state of the lunar interior was obtained thanks to the lunar seismic experiment. It was carried out as part of the Apollo program and consisted of continuous recording of natural moonquakes and meteorite falls using seismometers installed by American astronauts on the lunar surface and also from active research using artificial sources of seismic waves: explosions and impacts when the spacecraft was falling on the surface of the Moon [3, 4]. The first lunar seismometer was installed on July 20, 1969, in the Sea of Tranquility on the visible side of the Moon during the Apollo 11 expedition. It was powered by solar panels and kept a seismic record only during the day at the period from November 1969 to December 1972. The visible side of the Moon created a network of four automatic seismic stations of the same type, which were powered by isotope power plants designed for 10 years of continuous operation. Each of these stations was equipped with two types of seismometers and an electronic unit. Three long-period seismometers are arranged vertically and horizontally and allow to record seismic waves from distant sources and determine the direction to the epicenter. For a more accurate separation of waves at short distances, a vertical short-period seismometer was used. These seismometers felt negligible displacement of the surface, corresponding in the size of the atom (10–8 cm). On Earth, it is almost impossible to observe with such sensitivity of equipment: microseisms, ocean waves, wind, and industrial mechanism noise. The Moon, due to the absence of the atmosphere, hydrosphere, biosphere, and active internal processes, is an ideal testing ground for ultraprecise seismic surveys.

Lunar seismograms are completely different from those on Earth: the amplitude of recorded seismic oscillations increases gradually and decreases even more slowly (the "seismic sound" on the Moon lasts for hours). Records on the vertical and horizontal seismometers are not similar to each other, and there are no clear wave arrivals in the subsequent part of the recording—to isolate them; one has to resort to various methods: use narrowband frequency, polarization filters, etc. For these initially fascinate features of the lunar seismograms, an explanation has been received. The fact is that the upper layer of the Moon is very heterogeneous and broken by cracks due to meteorite falling. Seismic signals, scattered on these heterogeneities, stretch "in time" and are destroyed. At the same time, due to the absence of air and water on the Moon, the heat loss of elastic energy is not large, and, conversely, the so-called seismic quality factor is large—therefore, the oscillations do not damp out for a very long time. The features of seismic wave propagation in such a heterogeneous medium are fairly accurately described by the diffusion theory equations. Using these equations, the following estimates of the properties of the lunar scattering layer were obtained: its effective thickness is 15–25 km, and the size of heterogeneities is 2–5 km (and the degree of heterogeneity decreases with depth). Among the natural seismic events on the Moon emit tidal, thermal, and tectonic moonquakes, as well as the fall of meteorites. Seismologists have developed criteria for recognizing the nature of an event by the type of its record, similar to those used on Earth to identify earthquakes and nuclear explosions.

The absolute majority of all seismic phenomena on the Moon (about 90%) are relatively rare, regular, very weak tidal moonquakes with very deep focuses. At present, seismic data obtained using lunar seismometers for the period from November 1969 to July 1972 have been analyzed. During this time, about 5400 seismograms of tidal moonquakes were recorded. On average, about 600 such moonquakes are recorded annually in the Ocean of Storms at the Apollo 12 station, 650 in the Apennines (Apollo 15), 1500 in the region of Fra Mauro ("Apollo 14"), and about 3000 in the area of the Descartes crater ("Apollo 16"). The difference in the number of similar seismic phenomena recorded by the stations is explained by the ground conditions: there are more moonquakes where the upper layer of the soil and the underlying elastic rocks—breccias—are more powerful. All stations recorded an average of about 700 tidal moonquakes per year. All tidal moonquakes are very weak. The energy released in their focuses is 10^7–10^9 erg. The conditional magnitude of the event, adopted in seismology—the magnitude—for tidal moonquakes is 0.5–1.3, whereas for the strongest earthquakes, it reached 8.5. On Earth, earthquakes, as weak as tidal moonquakes, cannot be distinguished against the background of more intense microseisms. If we sum up the energy of all known tidal moonquake sources and assume that the seismicity on the reverse side is not higher than in the Descartes region, then we get the total seismic energy consumption of tidal moonquakes per year equal to 10^{10}–10^{12} erg. This is more than trillions of times less than earthquake energy. The curves of the number of moonquakes and the amplitude of ground oscillations in a seismic wave, equivalent to the "frequency patterns" of earthquakes, show the strength of the greatest tremors, as well as the number of the weakest shocks. The slope for moonquakes curve is 1.5–3.7 times steeper than for tectonic earthquakes. This means that the Moon is designed in such a way that many weak seismic shocks occur in it and strong ones are impossible. Similar dependences on Earth are obtained for small-scale volcanic earthquakes, as well as for underwater earthquakes in the regions of the middle oceanic ridges. Due to the fact that the seismic stations on the Moon are located at the vertices of a triangle with sides of about 1000 km, it is possible to determine both the position of the epicenter and the depth of the moonquake source. However, due to the small number of stations, the accuracy of such determinations is small (50–100 km). For all focuses, the form of recording remains constant throughout the observation period. According to this property, 41 focal zones were allocated, coordinates of epicenters were determined for 27 of them, and focal depths for 18 of them. The epicenters are grouped into two narrow extended seismic belts. The first is located approximately along the meridian (20–30° W longitude); it starts at 30° south latitude and stretches 2000 km to 40° S. latitude (expanding from 100 km in the north to 200–300 km in the south). The second belt, which is more than 300 km wide, stretches for 1800 km from the center of the visible side of the Moon to the east-northeast. Near its continuation there is a single epicenter fixed on the far side of the Moon. Some connection between the position of the epicenters and the features of the lunar relief is planned. The epicenters of the first seismic belt, comprising half of the focus and 63% of the total number of moonquakes, run along the western edge of the Seas of Rains, Cognized, and Clouds (and even slightly deviate to the east along the contour of the Sea of Clouds). At the same time, 50% of the number of moonquakes occurs on a 700-km stretch in the area of the junction of the Ocean of Storms with the Seas Cognized and the Clouds. More than half of the epicenters of the second seismic belt also fall to the margins of the lunar seas: two are in the south of the Sea of Clarity, two are in the north and east of the Sea of Crises, and three are in the mountain regions separating the Seas of Clarity and Tranquility. All tidal moonquakes occur deep in the depths of the Moon. This is evidenced by clear phases of transverse waves, large amplitude of the first arrivals, a shorter rise time of amplitudes than seismic phenomena of impact origin, and a relatively simple waveform in the interval between the arrivals of longitudinal and transverse waves. Focuses are located at depths of 400–900 km; however, >85% of them fall within a narrower interval—600–800 km. The stability of the form of records of moonquakes from one source (for one station) means that the size of the source zone is no more than 10–20 km. The zone of the epicenter of tidal moonquakes has a complex relief. The rise of focuses in the northern, southern, and central parts of the first seismic zone is planned. In the second zone, the deepest foci are located, and the western half has great depths. In general, the maximum depths of the focuses fall on the equatorial region (including the source on the far side of the Moon), while the minimum depths correspond to the central part of the southwest quarter of the visible side of the Moon [8].

A strict order is observed in the sequence of tidal lunar storms—an increase in the number of moonquakes after 13, 27, and 206 Earth days. An even longer (6-year) variability in the number of tidal moonquakes, which managed to manifest itself only in the longest series of observations at the Apollo 12 station, is also assumed. The maximum number of moonquakes occurred in the first half of 1970, then a smooth decrease in activity was observed, and in 1972, the Moon was the most passive. The next rise in seismicity was expected in 1976. The observed periodicity of moonquakes can be decoded based on the laws of rotation and motion of the Moon in the gravitational fields of the Earth and the Sun [9]. The orbit of the Moon is hardly one of the most complex planetary orbits. The Moon rotates around the Earth in an ellipse with an average distance of 384 thousand km and with an average eccentricity of the orbit of 0.05. The period of its circulation is 27 and 32 Earth days, during the same time the full rotation of the Moon from west to east around the axis takes place. The lunar equator has a slight slope to the ecliptic and lunar orbit. In the system, Earth-Moon tidal forces act. Viscosity of the planets effects on the dissipation of tidal energy. In addition, the gravity of the Sun causes periodic variations in the lunar orbit, manifested in changes in the eccentricity and distance of the Earth-Moon in the perigee. As a result of all this, a characteristic feature arises—the physical libration of the Moon, which complicates its movement and rotation [11]. It has a latitudinal component with a period of 6 years and a longitude of 206 days.

Initially, when studying tidal moonquakes, it seemed that their tremors were timed to the moments of the apogee and perigee of the Moon in the orbit around the Earth. As data accumulated, it turned out that the picture is more complicated: the peaks of seismic activity are shifted in accordance with the periods of the Moon's libration. At the same time, the seven-month maximum of seismic activity is tied to the greatest eccentricity of the lunar orbit.

Based on the above, none of the researchers of the seismicity of the Moon now doubts the external, cosmic nature of tidal moonquakes—the role of the "trigger mechanism" in them is played by the forces of gravity of the Earth and the Sun. The problem of forecasting planetary shocks, so complex and important on Earth, on the Moon is solved simply—the "schedule" of tidal moonquakes can be made on the basis of the laws of celestial mechanics. For example, the dependence of the moment of seismic shock in the corresponding epicenter on the position on the Moon's surface of a point lying on the straight line connecting the centers of mass of the Earth and the Moon is noted. So, tidal moonquakes can be predicted, but is it so important? After all, they are extremely weak and harmless, and, in particular, future designers of lunar cities and rocket tracks will not need to introduce corrections for the seismic resistance of structures into their calculations.

Tectonic moonquakes. The seismic experiment on the Moon was mainly focused on the detection of tectonic moonquakes; therefore, the stations were established in the contact areas of large-scale surface structures. However, for the entire time of observations, only 11 shocks were recorded, possibly of a tectonic nature. But despite their small amount, they raise the total seismic energy of moonquakes by several orders of magnitude (up to 10^{15} erg). The first characteristic feature of such events is the high frequency of their records (according to this feature, they differ sharply from both tidal moonquakes and meteorite strikes). The arrivals of the longitudinal and transverse waves are very clear, which indicates a small scattering of waves near the source. The slope of the "repeatability curves" is much smaller than that of tidal and thermal moonquakes and closer to earthquakes. The energy of tectonic moonquakes is several orders of magnitude higher than that of tidal moonquakes; their magnitude reaches 4. All detected tectonic moonquakes were out of the network of lunar seismic stations at distances greater than 600 km: 10 out of 11 were recorded by all stations, and for 4 out of 11, only the azimuth and distance were determined, for 7 events (with an accuracy of 5°). All the epicenters are located on the periphery of the border of the visible and reverse sides of the moon: 9—on the visible and 2—on the back (without an explicit connection with the surface structures). Tidal tectonic moonquakes are not found in the southeast quarter of the visible side of the moon. They are not regular; their form of recording is not repeated. The depth of the foci is not determined precisely; according to the nature of the recording of one of the strong tectonic moonquakes, the lower estimate of the depth of its source—300 km—was obtained, while the foci are deeper than the scattering layer of the crust with a thickness of 25 km.

During the Apollo expeditions, seismic studies of the structure of the subsoil of various scales were carried out: seismic exploration of the upper part of the section, sounding of the crust, and seismic scanning of the mantle. The first detailed study of the velocity of longitudinal waves in the lunar soil was carried out in the Fra Mauro area during the Apollo 14 expedition. Three seismometers recorded astronauts tapping on the ground at a distance of 100 m. More distant points were obtained during the takeoff of the lunar cabin and with the help of special grenades that were blown up after the astronauts departed on command from Earth. A similar experiment was conducted by Apollo 16 astronauts in the continental region of the Descartes crater. Depths of up to 200 m were studied. Finally, during the Apollo 17 expedition in the area of Taurus-Littrow, it was possible to "light up" the structure of the upper layers to a depth of 1.5 km. According to the records at the seismological stations of the fall of the spacecraft, the propagation speeds of transverse seismic waves were determined. As a result of all studies, it was established that the upper part of the Moon consists of separate layers in which the speed of seismic waves increases abruptly with depth. The upper two layers have properties similar in different regions, distant from each other and folded in different breeds. This is due to their identical origin: the scatter of debris during the impact of large meteorites and their subsequent crushing by small ones. Characteristics of lunar soil (regolith): power (2–12 m), velocity of longitudinal waves (90–115 m/s), velocity of transverse waves (35–37 m/s), ratio of velocities of longitudinal and transverse waves (2.7–2.9), Poisson's ratio [2] (0.42–0.43), density (1.5–1.6 g/cm^3), and porosity (more than 50%). In the layer of detrital material (breccias), the power is 18–38 m, the velocity of the longitudinal waves is 300 ± 50 m/s, the ratio of the velocities is 2.2–2.4, the Poisson's ratio is 0.37–0.40, and the density is 1.7–1.8 g/cm^3. Compared with the Earth's soil, the Moon has very low velocities and a high ratio of velocities (a large Poisson's ratio).

2.1 Probing of the crust

The lunar crust was studied in the southeastern part of the Ocean of Storms, where the Apollo 12 and 14 seismic stations were located 180 km away. In this case, a rather unusual (for the Earth) method of exciting elastic waves was used—dropping the spent compartments of spaceships. The Apollo expeditions produced two seismic shocks on the surface of the moon. The first shock was, when the third stage of the Saturn-5 rocket weighing more than 14 tons was sent on command from the Earth to a given point on the surface of the Moon. With an almost vertical fall at a velocity of 2.5 km/s at the surface, seismic waves of such force were excited as in the explosion of 10 tons of TNT. The second shock was caused when the takeoff stage, after the crew returned to the main compartment, was dropped near the corresponding seismic station. With a mass of 2.4 tons, the velocity at the surface of 1.7 km/s, and the "gentle" approach trajectory to the surface, the impact of the lunar cabin was equivalent to an explosion of 800 kg of TNT. The network of lunar seismic stations recorded nine such "artificial" impacts, as well as one natural attack from a meteorite weighing more than 1 ton, which "successfully" fell (May 13, 1972) 140 km north of the seismic station Apollo 14. A total of 14 seismograms were received, 9 of them at a source-receiver distance of 67–358 km and 5 at a distance of 850–1100 km. Seismic records were analyzed in accordance with the methods adopted in the practice of terrestrial deep seismic sounding of the crust. Due to the insufficient accuracy of observations and errors in determining the arrival time of the waves on the seismograms, the velocity structure of the lunar crust was determined with some approximation. In the upper third of the crust, the velocity of longitudinal seismic waves rapidly increased from values of 100 m/s in regolith to 4.5–5.0 km/s at a depth of 10 km and 5.5–5.8 km/s at a depth of 20 km. In the lower part of the crust, the velocity of seismic waves remains almost stable—7.0 km/s. However, only apparent stability is possible, because the lower layers of the crust are "illuminated" so inaccurately even on more detailed observations on Earth that it is sometimes difficult to distinguish a slight increase in the velocity of seismic waves from its gradual decrease. In the mantle of the Moon, the velocity of seismic waves increases abruptly to values of 8 km/s and more. The boundary of the crust-mantle, analogous to the boundary of Mohorovich on Earth, has not yet received on the Moon. The nature of the transition from the crust to the mantle remained unclear, until analysis of their amplitudes and spectra was added to the analysis of the travel times of the waves. The bursts of waves in the subsequent part of the record, identified as reflections from the crust-mantle boundary, showed that this is, in fact, not a boundary, but a transition layer with a thickness from 3–4 to 10–12 km. In some interpretations, a sharp increase in the velocity inside the crust from 5.8 to 6.8 km/s at a depth of 25 km is assumed. In the future, we will be interested in the state of the upper part of the lunar crust, so in the review we omit the results obtained by the structure of the mantle and the core of the Moon. Seismic observations make it possible to determine not only the propagation velocity of longitudinal and transverse waves but also a measure of the proximity of real matter to the model of ideal elasticity—the degree of energy absorption due to irreversible heat losses along the path of the seismic waves. The seismic quality of lunar rocks was estimated in several ways: laboratory (measurements on samples placed in a vacuum), theoretical (comparison of seismograms with calculations), and experimental (measurement of the law of decreasing amplitudes on seismograms). In the latter method, attenuation of the amplitudes of seismic recording was considered depending on time, distance, and frequency. Different definitions gave consistent results. According to the new data, the layered structure of the Moon is distinguished by the seismic quality factor. In the upper layer (regolith), the quality factor for longitudinal waves is 100–300, and in the scattering layer of the crust—3000–5000, in the entire 500-km thickness of the lithosphere—5000 (according to some definitions—7000–10,000), deeper the quality factor drops to 3500 (at a depth of 600 km), 1400 (950 km), and 1100 (1200 km). In the asthenosphere of the moon, the Q on longitudinal waves does not exceed 500. According to the amplitudes of the transverse waves, the seismic Q of the upper 300 km sequence is estimated at 4000; in the 500–800 km layer, it decreases to 1500, and in the asthenosphere, it drops another 10–15 times.

2.2 Lunar mascons

A detailed study of the field of gravity of the Moon became possible after the launch of space satellites into the orbit of artificial satellites of the moon. Observations of satellite orbits were carried out with the help of three ground stations. The first constructions of the picture of the gravitational field of the Moon were carried out by Soviet researchers on the results of the flight of the Luna-10 spacecraft, and further data were updated on observations of the orbits of artificial satellites of the Lunar Orbiter series, as well as on those sections of the Apollo spacecraft routes, their orbits around the Moon were determined only by its field of gravity. The gravitational field of the Moon turned out to be more complicated and heterogeneous than the Earth, the surface of equal potential of gravity is more uneven, and the sources of anomalies are located closer to the surface of the moon. A significant feature of the lunar field of gravity was the large positive anomalies associated with round seas, which were called mascons (from the English—"mass concentration"). When approaching the mascon, the satellite velocity increases; after passage, the satellite slows down slightly, with the orbit altitude changing by 60–100 m. At first, mascons were discovered in the seas of the visible side: Rains, Clarity, Crises, Nectar, and Humidity; their sizes reached 50–200 km (they fit into the contours of the seas), and the anomaly values were 100–200 mgal [12]. The anomaly of the Sea of Rains corresponded to an excess mass of about (1.5–4.5) $\times 10^{-5}$ the masses of the entire Moon. Subsequently, more massive mascons were discovered on the border of the visible and reverse sides in the Eastern and Regional Seas, as well as a huge mascon in the equatorial zone of the center of the far side of the Moon. There is no sea in this place; therefore, the mascon is called "Hidden." Its diameter is more than 1000 km; the mass is five times the excess mass of the Sea of Rains. Hidden mascon is able to deflect a satellite flying at an altitude of 100 $\times$ 1 km. The total excess mass corresponding to the positive anomalies of gravity exceeds 10^{-4} the mass of the moon. A number of negative anomalies were associated with the lunar mountains: the Jura, the Caucasus, the Taurus, and the Altai. Anomalies of gravity reflect the distribution of mass of matter in the depths of the Moon. If, for example, we assume that mascons are created by point masses, then their depths should be about 200 km in the Sea of Rains, 280 km in the Sea of Clarity, 160 km of Crises, 180 km of Calmness, 100 km of Plenty, 80 km of Sea of knowledge, and 60 km of Ocean of Storms. Thus, gravity measurements found a nonuniform density distribution in the upper mantle.

2.3 Electrical conductivity

None of the lunar expeditions did directly measure the electric field of the moon. It was calculated from the variations of the magnetic field recorded by magnetometers at the Apollo 12, Apollo 15, Apollo 16, and Lunokhod-2 stations. The Moon is devoided of the magnetosphere, but during the magnetosphere, with its rotation around the Earth periodically turns out to be in full Moon in the unperturbed Earth magnetosphere, in the new Moon in the solar wind, and twice for 2 days in the transitional one-shock layer. Fluctuations of an external interplanetary magnetic field penetrate the Moon and induce an eddy current field in it. The rise time of the induced field depends on the distribution of electrical conductivity in the lunar interior. Simultaneous measurements of the external alternating field over the Moon and the secondary field on the surface allow us to calculate the lunar conductivity.

The Moon is arranged "conveniently" for magnetic-telluric sounding. The interplanetary magnetic field, extended from the Sun, is uniform; its front can be considered flat, and therefore for research, as on Earth, a network of laboratories is not necessary. Due to the fact that the Moon has a higher electrical resistance than the Earth, two hourly observations are enough for its sensing, whereas annual Earth winds are needed on Earth. The solar wind flowing around the Moon, having high conductivity, seems to envelop the Moon without letting it out to the surface fields. Therefore, on the solar side of the Moon, only the horizontal component of the alternating magnetic field can be used, whereas on the night side, where the vertical component works, the situation is more similar to that of the Earth. The Apollo magnetometers recorded the Moon's response in the solar wind on the night and day sides, and also in the geomagnetic plume, where the plasma effects of the solar wind are minimized. In the crater of Lemonier on the solar side of the Moon, on the Lunokhod-2, the solar fluctuation magnetic field was registered. In this case, the horizontal component of the magnetic field reflects the depth electrical conductivity of the Moon, and the magnitude of the vertical component for a long time characterized the intensity of the external field of the Moon. An experimental graph of apparent resistivity was interpreted by comparing with theoretical curves. Soviet (Vanyan et al. [13]) and foreign researchers constructed various models of the electrical conductivity of the Moon. Differing in some details, they give generally similar distributions of the electrical properties of lunar material with depth: in the upper 200 km, there is a poorly conducting layer with a specific resistance of more than 10^6 ohm m; a layer of reduced resistance (10^3 ohm m) with a thickness of 150–200 km is deposited deeper, and the resistance increases by an order of magnitude to 600 km and then decreases again to 10^3 ohm m at a depth of 800 km.

The electrical sensing of the Moon carried out so far reveals the following main features: The Moon as a whole has a higher resistance than the Earth. On top of it is a powerful insulating layer; with depth the electrical conductivity grows. The radial separation of the Moon was found, and a heterogeneity in the horizontal direction was noted with respect to electrical resistance. The temperature inside the Moon was estimated for different compositions of the mantle using the electrical conductivity profiles and the dependence of conductivity on temperature. In all cases, to a depth of 600–700 km, the temperature lies below the melting point of basalt and at great depths reaches or exceeds it.

2.4 Moon's crust

Like on Earth, the Moon has a crust separated from the mantle by a sharp boundary. The thickness of the lunar crust in the southeastern part of the Ocean of Storms (60–65 km) is the same as in the Pamir or Himalayas and more than just oceanic (7–10 km) but also continental crust (40 km). The lunar crust is one-thirtieth of the size of the moon, and thus, in relation to the radius of the planet, it is five times thicker than the average crust. Seismic measurements, which give the most accurate estimates of crustal thickness, have so far been carried out only in the Ocean of Storms. According to other, especially gravimetric, data, we can conclude that the thickness of the crust in different areas is different: in the eastern hemisphere, as well as on the reverse side of the moon, the crust is several times more powerful than in the western. It is possible that, in the region of the Mascon Sea of Crises and Clarity, the denser subcrustal matter lies closer to the surface, here the thickness of the crust decreases to 70–80 km. The difference in the physical properties of the crustal rocks in different areas is noted not only in terms of seismic wave velocities and rock densities—they are differently magnetized and have different electrical conductivities.

Horizontal heterogeneity of densities leads to stresses that cause tectonic moonquakes at depths of 25–300 km. These stresses (100–200 kg/cm^2) are tens of times smaller than the horizontal forces that determine the tectonic activity of the Earth's lithosphere; therefore, tectonic moonquakes are so weak compared to earthquakes. In the light of new knowledge about the deep structure of the Moon, the picture of the preparation of moonquakes looks like this. Under the action of the forces of gravity of the Earth and the Sun, large tidal voltage drops occur in the moon. They concentrate on the contact of the hard outer and heated inner zones of the moon. This contributes to a complex, contrasting relief of the transition zone. Perhaps, the position of the epicenters of moonquakes reflects the direction of convective flows of matter in the asthenosphere. At the moments of increasing the attraction of the Moon by the Earth and the Sun, hot fluids and gases are injected into the transitional zone by pulses. They form a kind of "lubricant," which further facilitates the movement of blocks along the gap at the time of the moonquake. The dimensions of the foci, the intervals between the tremors, and their energy agree quite well within the framework of a theory describing the process of an earthquake as a rapid "ripping up" of cracks in weakened zones. On the Moon, breaks occur within homogeneous blocks of poorly cemented material. Therefore, from push to push, the shape of oscillations in the waves from each source is so well preserved. Because of the small size of the blocks, the shaking is not great. And their "schedule" is fully regulated by the gravitational "pointer" of the Earth and the Sun. There is not enough time for stresses to accumulate, as the next stress impulse and "lubrication" from the asthenosphere arrive—a weak moonquake occurs. Tidal forces of the Earth make the Moon shake often and weakly, not allowing it to accumulate forces for a powerful push.

In the article [14], in order to explain the reason for the manifestation of earthquakes of an "explosive" nature, a model of redistribution of lithostatic pressure during the formation of arch (anticlinal) structures of the Earth's crust was proposed. It is shown that in the castle part of the formed arch structure, the lithostatic pressure increases significantly, and this leads to a sharp decrease in the volume of the rock ("collapse"). The under locked part of the lithostatic pressure is removed, which leads to mechanical destruction of the rock ("explosion"). It was concluded that the processes of "collapse" and "explosive" destruction of rocks occur during the formation of arch structures and are a possible cause of the manifestation of earthquakes of "explosive" nature. When conducting electromagnetic monitoring in shock-hazardous rock massifs, which were as a result of mining under intense explosive effects in local instability zones, in the section, disintegration zones were found, united by morphology like a vault structure. This may indicate a dynamic hierarchical similarity of the response of the geological environment to the preparation of dynamic manifestations in the form of earthquakes or rock bursts.

3. Informative signs of the preparation of high-energy dynamic phenomena according to the mine seismological monitoring

In order to create a dynamic model adequate to the processes of preparing high-energy processes in the rock massive that are under a strong technogenic impact, it was necessary to use monitoring data in natural occurrence. For this, an analysis of the detailed seismic catalog of the Tashtagol underground mine for 2 years of observations from January 2006 to January 2008 was carried out. The data used are the space-time coordinates of all dynamic phenomena and massive responses that occurred within the mine field during this period, and explosions produced for mining the massive, as well as the values of the explosion energy and massive responses recorded by the seismic station [15]. The whole mine field was divided into two halves: the northwestern section and the west and new capital shaft areas, and the holes from 0 to 13 were designated by us as the northern section. The holes from 14 to 31, the southern ventilation and field drifts, the shaft of the south mine, and the development of the southeastern section are designated as the southern section. All events-responses with horizons with marks −140 m, −210 m, −280 m, and −350 m (maximum depth 800 m) were analyzed. Impacts in the form of explosions were carried out in the southern, southeastern, northwestern, and northern areas. The seismological catalog was also divided into two parts: the northern and southern, according to the events, responses, and explosions that occurred in the northern and southern parts of the mine field.

Phase portraits of the state of the massive of the northern and southern sections are constructed in coordinates E0(t) and d(E0(t))/dt, t—time, expressed in fractions of 24 h and E0—massive seismic energy in J. In the paper [15], morphology of the phase trajectories of the seismic response to explosive effects at various successive intervals of the southern section of the mine is analyzed. During this period, according to the data on the technological and mass explosions produced, most of the energy was pumped into the southern section of the mine. In addition, at the end of the year 2007, it occurred in the southern section, the one of the strongest rock bursts in the entire history of the mine. As a result of the analysis, the characteristic morphology of the phase trajectories of the massive response that is locally in time in a stable state is highlighted. On the phase plane, there is a local area in the form of a coil of interlaced trajectories and small emissions from this coil, not exceeding in energy the values of 10^5 J. At some intervals, this release exceeds 10^5 J, reaching 10^6 J and even 10^9 J [15, 16]. Obviously, there are two interdependent processes: the process of energy accumulation, which is reflected in the region attracting phase trajectories, and the process of resonant discharge of accumulated energy. It is interesting to note that after this reset, the system returns again to the same region attracting phase trajectories. This is confirmed by the detailed analysis of the phase trajectories of the seismic response of the massive before and after the strongest rock burst. However, the process of changing the state of the massive is strongly influenced by the process of a fairly regular external influence in the form of explosions of different powers. During the time between the explosions, the massive does not have time to allocate the energy received by it, which leads to a response delay and nonlinearity of its manifestation, which makes it difficult to predict the time of a highly energetic destructive event [17].

Based on the ideas presented in [18], the analyzed database was supplemented with data of spatial coordinates of explosions. On this basis, a new algorithm for processing seismological information of a detailed mine catalog was developed taking into account the kinematic and dynamic characteristics of deformation waves propagating at different velocities in the rock massif under intense external influence in the form of mass or technological explosions [19]. It was found that waves propagating at velocities from 10 to 1 m/h are the primary carrier of energy in the array and contribute to its release. Events occurring in the array with these speeds and having energy emissions less than 10^4 J contribute to the creep reorganization of the hierarchical inclusions of block parts of the array, which leads to the organization of a new section of dynamic instability. Events occurring in the array with these speeds and having an energy release greater than 10^5 J can be used as precursors which should be taken into account when adjusting the product of explosions in one or another part of the array. The complete absence of these events indicates an increase in the stress state in the massif of the mine as a whole.

3.1 Algorithm for processing seismological information to determine the informative features of the preparation of high-energy dynamic phenomena

In the present work, quantitative estimates have been made of the lag parameter of the high-energy response of the array to a number of anthropogenic impact, during which a significant part of the time was the absence of response of the array. Shock (Sh.36) with an energy of 8.14E+08 J happened at 25.11.2007 with coordinates x = 11,928 m, y = 11,627 m, and z = −264 m (+(−450 m)). It is marked, like all the rest of our studied responses, which occurred in the southern part of the mine with the letters Sh. and number 36. Explosions are designated as (i), where i is the number of the explosion for the period 2006–2008. We have obtained additional estimates of the distances from the explosion point to the response point of the massif. The coordinates of the explosions and massif responses are taken from the seismic mine catalog of the Tashtagol mine.

3.2 Discussions

As follows from the analysis of the data in **Tables 1–4** and **Figures 1–4**, the response of the massif Sh.36 in the form of a high-energy response only appears, starting with the distances between the point of impact and the response from 100 m to 200 m. At the same time, the lag time of the massif response to the impact occurred in the form of a rock shock is tens and even hundreds of days. Therefore, despite the fact that the Sh.36 response from the impact (78) (**Table 4**) occurred almost instantly,

Imp.-Sh.	dx	dy	dz	r	dt	Eimp	E0
(1)-Sh.1	0	−23	10	25	77	5.40E+04	2.7E+06
(1)-Sh.2	−18	−1	3	18	111	5.40E+04	1.48E+04
(23)-Sh.7	−24	−16	3	29	0.3	2.15E+07	1.56E+04
Sh.10–Sh.17	−20	17	−18	32	40	1.17E+04	1.56E+04
Sh.10–Sh.18	−21	−31	−13	40	48	1.17E+04	1.56E+04
(38)-Sh.17	−30	−24	−1	38	28	7.97E+03	1.56E+04
(46)-Sh.21	−25	11	3	27	2	9.52E+07	7.48E+04
Sh.31–Sh.32	1	−2	−3	4	0.3	1.10E+04	1.04E+04

Imp, impact; Sh, response; dx, dy, dz, difference of Cartesian coordinates of the points of impact and response in m; $r = \sqrt{dx^2 + dy^2 + dz^2}$, *the distance (m) between the point of impact and the point of response; dt, time difference of impact and response points in 24 h; Eimp, impact energy (J); E0, energy response (J).*

Table 1.
The impact and the response of the mine southern part massif within the distance from 0 m to 50 m from the points of explosions.

Imp.-Sh.	dx	dy	dz	r	dt	Eimp	E0
(23)-Sh.5	−56	−34	28	71	0.03	2.15E+07	1.24E+04
(30)-Sh.9	55	13	−13	58	18	6.35E+06	1.65E+04
(34)-Sh.10	43	−49	−58	87	2	2.44E+04	1.17E+04
(38)-Sh.11	4	−41	41	58	15	7.97E+03	1.04E+04
(38)-Sh.18	−31	−72	4	78	36	7.97E+03	1.17E+04
Sh.11-Sh.16	26	62	6	67	11	1.04E+04	1.10E+04
(42)-Sh.13	24	−24	−59	68	0.02	1.19E+09	4.04E+04
Sh.15-Sh.19	75	60	18	98	22	2.92E+04	2.44E+04
Sh.23-Sh.28	−7	−62	−39	74	138	2.92E+04	1.16E+04
Sh.25-Sh.27	42	42	−40	72	38	3.74E+04	4.04E+04
(72)-Sh.34	72	54	22	93	42	1.80E+06	8.68E+04
(78)-Sh.35	97	18	7	99	0.1	8.68E+04	2.67E+04

Imp, impact; Sh, response; dx, dy, dz, difference of Cartesian coordinates of the points of impact and response in m; $r = \sqrt{dx^2 + dy^2 + dz^2}$, *the distance (m) between the point of impact and the point of response; dt, time difference of impact and response points in 24 h; Eimp, impact energy (J); E0, energy response (J).*

Table 2.
The impact and the response of the mine southern part massif within the distance from 50 m to 100 m from the points of the impacts.

Imp.-Sh.	dx	dy	dz	r	dt	Eimp	E0
(16′)-Sh.36	111	97	−22	149	575	1.24E+04	8.14E+08
(21)-Sh.36	117	53	−34	133	539	1.65E+04	8.14E+08
(23)-Sh.4	69	−82	19	109	0.01	2.15E+07	1.92E+04
(23)-Sh.8	−28	−121	−2	124	0.06	2.15E+07	3.74E+04
(35)-Sh.36	118	61	34	137	400	1.92E+04	8.14E+08
(38)-Sh.36	77	86	−18	117	393	7.97E+03	8.14E+08
(39)-Sh.36	100	45	33	114	392	2.07E+06	8.14E+08
(41)-Sh.36	102	74	41	132	378	3.95E+06	8.14E+08
Sh.4-Sh.5	−125	48	9	134	0.03	1.92E+04	1.24E+04
Sh.5-Sh.6	−35	−118	−13	124	0.96	1.24E+04	3.35E+05
Sh.7-Sh.8	4	−105	5	105	0.09	1.04E+04	3.74E+04
Sh.20-Sh.21	−95	−28	−81	128	1	6.60E+04	7.48E+04
(43)-Sh.18	13	−88	−46	100	7	1.55E+06	1.17E+04
(43)-Sh.36	121	70	24	142	364	1.55E+06	8.14E+08
(44)-Sh.18	32	−108	17	114	0.08	3.98E+06	1.17E+04
(68)-Sh.33	31	−137	−20	142	70	2.70E+06	3.48E+05
(68)-Sh.36	89	70	34	118	70	2.70E+06	8.14E+08
(70)-Sh.34	106	33	17	112	56	2.06E+06	8.68E+04
(71)-Sh.36	63	75	48	109	49	1.65E+05	8.14E+08
(72)-Sh.36	55	91	39	113	42	1.80E+06	8.14E+08
(78)-Sh.36	80	55	44	107	0.06	8.68E+04	8.14E+08

Imp, impact; Sh, response; dx, dy, dz, difference of Cartesian coordinates of the points of impact and response in m; $r = \sqrt{dx^2 + dy^2 + dz^2}$, the distance (m) between the point of impact and the point of response; dt, time difference of impact and response points in 24 h; Eimp, impact energy (J); E0, energy response (J).

Table 3.
The impact and the response of the mine southern part massif within the distance from 100 m to 150 m from the points of the impacts.

it was preceded by a long process of preparing a resonant energy release [22]. Based on the ideas presented in [18], the analyzed database was supplemented with data of spatial coordinates of impacts. On this basis, a new algorithm for processing seismological information of a detailed mine catalog was developed taking into account the kinematic and dynamic characteristics of deformation waves propagating at different velocities in the rock massif under intense external influence in the form of mass or technological explosions [19]. It was found that waves propagating at speeds from 10 to 1 m/h are the primary carrier of energy in the massif and contribute to its release. Events occurring in the massif with these velocities and having energy emissions less than 10^4 J contribute to the creep reorganization of the hierarchical inclusions of the block parts of the massif, which lead to the organization of a new section of dynamic instability. Events occurring in the array with these speeds and having an energy release greater than 10^5 J can be used as precursors which should be taken into account when adjusting the product of explosions in one or another part of the array. The complete absence of these events indicates an increase in the stress state in the massif of the mine as a whole. In [20], an algorithm for processing seismological information of a detailed mine catalog was developed to determine the informative features of the preparation of high-energy dynamic phenomena. For this purpose, quantitative

Imp.-Sh.	dx	dy	dz	r	dt	Eimp	E0
(23)-Sh.6	−91	−152	15	178	1	2.154E+07	3.35E+05
(29)-Sh.36	170	91	−23	194	434	2.07E+06	8.14E+08
(30)-Sh.36	174	76	2	190	427	6.35E+06	8.14E+08
(34)-Sh.36	130	78	23	153	407	2.44E+04	8.14E+08
(40)-Sh.36	140	63	17	154	385	2.70E+06	8.14E+08
Sh.4–Sh.6	−160	−70	4	175	0.99	1.92E+04	3.35E+05
Sh.20–Sh.37	38	−146	14	152	336	3.35E+05	6.39E+04
(42)-Sh.14	−78	−180	26	198	0.06	1.19E+09	1.92E+04
(42)-Sh.15	163	−38	−31	170	2	1.19E+09	2.92E+04
Sh.13–Sh.15	187	−14	−28	190	2	1.48E+04	2.92E+04
(44)-Sh.36	140	50	50	157	357	3.98E+06	8.14E+08
(46)-Sh.36	140	64	−17	155	336	9.52E+07	8.14E+08
(59)-Sh.36	115	117	1	164	197	1.10E+04	8.14E+08
(74)-Sh.36	98	111	−54	158	35	2.93E+06	8.14E+08
(78)-Sh.33	1	−168	58	178	0.01	8.68E+04	3.48E+05

Imp, impact; Sh, response; dx, dy, dz, difference of Cartesian coordinates of the points of impact and response in m; $r = \sqrt{dx^2 + dy^2 + dz^2}$*, the distance (m) between the point of impact and the point of response; dt, time difference of impact and response points in 24 h; Eimp, impact energy (J); E0, energy response (J).*

Table 4.
The impact and the response of the mine southern part massif within the distance from 150 m to 200 m from the points of the impacts.

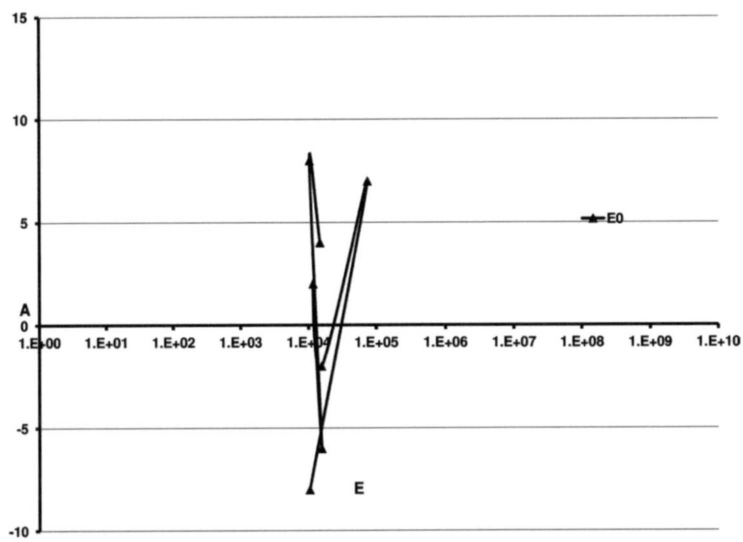

Figure 1.
Phase diagrams of the dynamical state of the massif southern mine part during the period 2006–2008 years, r = 0 m–50 m. Horizontal axe: E = Eo; vertical axe: $A = aLgf$, $f = \left|\frac{\partial E_V}{\partial t}\right|$, $a = \text{sign}\ \partial E_V$ $E_V = E_{imp} - Eo$.

estimates have been made of the lag parameter of the high-energy response of the array to a number of anthropogenic impacts, during which a significant part of the time was the absence of the response of the array. Response 8.14E+08 J happened 25.11.2007 year with coordinates x = 11,928 m, y = 11,627 m, and z = −264 m (+(−450 m)). We have obtained additional estimates of the distances from the impact

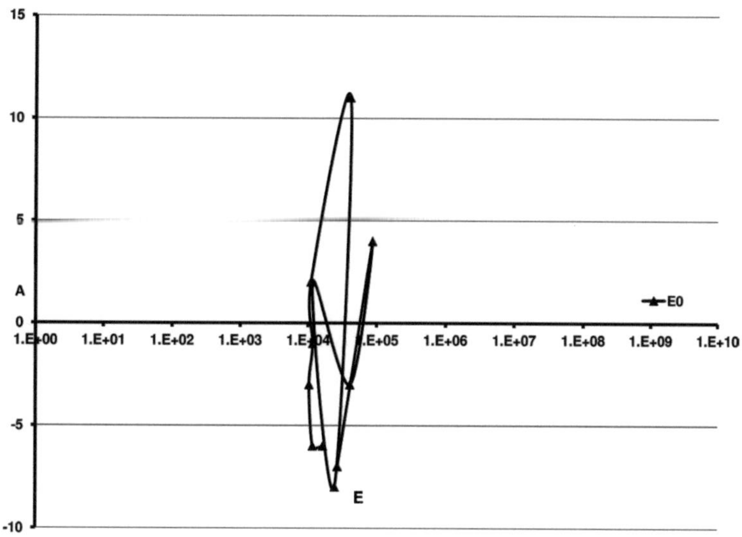

Figure 2.
Phase diagrams of the dynamical state of the massif southern mine part during the period 2006–2008 years, r = 50 m–100 m. Horizontal axe: E = Eo; vertical axe: $A = aLgf$, $f = \left|\frac{\partial E_V}{\partial t}\right|$, $a = sign\ \partial E_V$ $E_V = E_{imp} - Eo$.

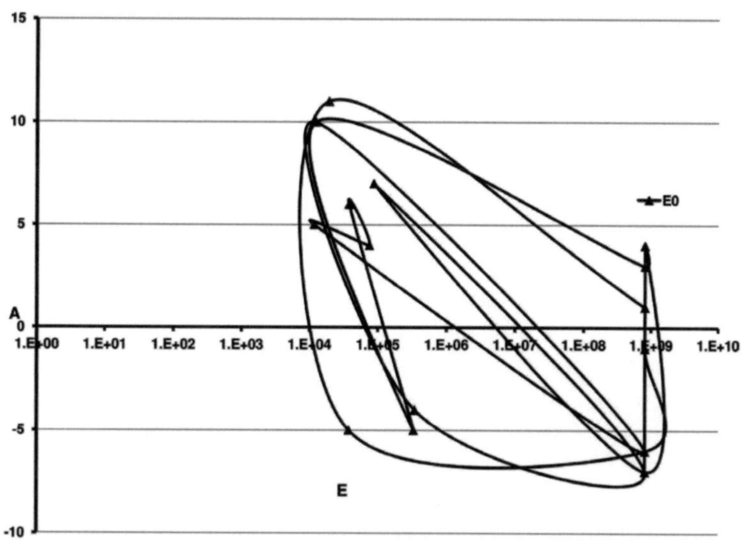

Figure 3.
Phase diagrams of the dynamical state of the massif southern mine part during the period 2006–2008 years, r = 100 m–150 m. Horizontal axe: E = Eo; vertical axe: $A = aLgf$, $f = \left|\frac{\partial E_V}{\partial t}\right|$, $a = sign\ \partial E_V$ $E_V = E_{imp} - Eo$.

point to the response point of the massif. The coordinates of the impacts and massif responses are taken from the seismic mine catalog of the Tashtagol mine. As follows from the analysis performed, we can repeat that the massif response in the form of a high-energy response is manifested only from the distances between the point of impact and the response from 100 m to 200 m. At the same time, the lag time of the array response to the impact occurred in the form of an explosion is tens and even hundreds of days. Therefore, despite the fact that the response from the last impact occurred almost instantly [20], it was preceded by a long process of preparing resonant energy release [21], which must be accompanied by electromagnetic monitoring of the occurrence and accumulation of disintegration zones in the massif volume: dx = 100–180 m, dy = 33–180 m, and z = (−210 − (−300)) + (−450) m. It is important

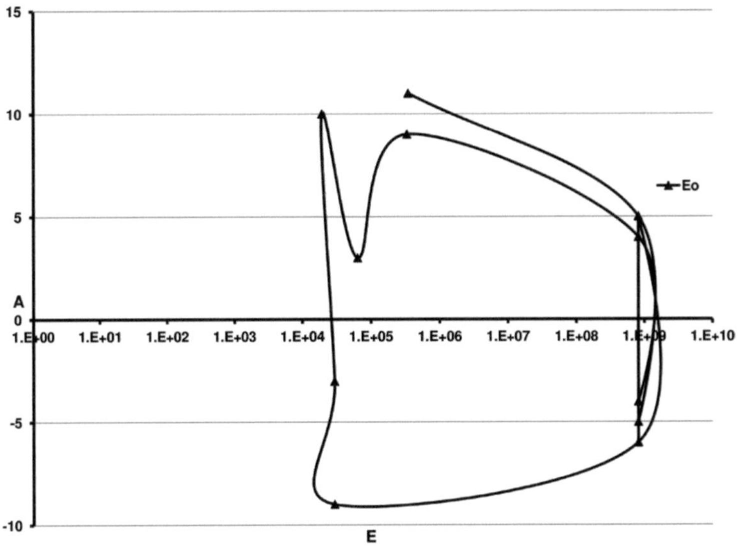

Figure 4.
Phase diagrams of the dynamical state of the massif southern mine part during the period 2006–2008 years, r = 150 m–200 m. Horizontal axe: E = Eo, vertical axe: $A = aLgf$, $f = \left|\frac{\partial E_V}{\partial t}\right|$, $a = sign\ \partial E_V$ $E_V = E_{imp} - Eo$.

to take into account not only the direct impact of explosions on this process but also the responses that appear after a while as shocks that contribute to the resonant release of energy. The developed algorithm for processing seismological information of the detailed mine catalog allows extracting additional important information for predicting hazardous phenomena in ore mines and for developing the theory of dynamic phenomena in natural geological, geophysical, and lunar environments.

As a result of the analysis performed in [23], the use of the theory of catastrophe recommendations [24] on the prediction of dynamic phenomena can be drawn from the following conclusions. The processes occurring in mountain massive are dynamic processes that can be controlled following the recommendations provided by the theory of catastrophes. In these processes, the energy values during explosions and the location of these explosions relative to the region of the massif under study or being processed are the control parameters. The internal parameters are the kinematic and dynamic parameters of deformation waves [21, 25–27], as well as the structural features of the massif through which these waves pass [22]. The use of analysis methods for short-term and medium-term forecasting of the state of a mountain massif only by using control parameters is not enough because the massif has sharp heterogeneities. However, the joint use of qualitative recommendations of the theory of catastrophes and the spatial-temporal data of changes in the internal parameters of the massif will allow preventing catastrophes in mountain terrestrial and lunar massive.

4. Conclusions

In the case of studying the state of the lunar mountain massive, it is necessary to organize active seismic and deformation monitoring to the similarity organized on the Earth within unstable mountain massive. Since the Moon is at a considerable distance from the Earth, the active effect as an excitation source can be of electromagnetic or laser type. The basic principle of monitoring should be active and regularly repeated, and then the processing algorithm described above can be used, bypassing the time paradox, to predict the state of the lunar mountain massive. At present, theoretical results on modeling the electromagnetic and seismic field in a layered medium with inclusions of a hierarchical structure are in demand. Simulation algorithms are constructed in the electromagnetic case for 3D heterogeneities, in the seismic case for 2D heterogeneities [28–30]. It is shown that with an increase in the degree of hierarchy of the environment, the degree of spatial nonlinearity of the distribution of the components of the seismic and electromagnetic fields increases, which corresponds to the detailed monitoring experiments conducted in the shock-hazardous mines of the Tashtagol mine and SUBR. The theory developed demonstrated how complex the process of integrating methods using an electromagnetic and seismic field to study the response of a medium with a hierarchical structure. This problem is inextricably linked with the formulation and solution of the inverse problem for the propagation of electromagnetic and seismic fields in such complex environments. In [31, 32], the problem of constructing an algorithm for solving an inverse problem using the equation of a theoretical inverse problem for the 2D Helmholtz equation was considered. Explicit equations of the theoretical inverse problem are written for the cases of electromagnetic field scattering (E and H polarization) and scattering of a linearly polarized elastic wave in a layered conducting and elastic medium with a hierarchical conducting or elastic inclusion, which are the basis for determining the contours of misaligned inclusions of the lth rank of the hierarchical structure. It is obvious that when solving the inverse problem, it is necessary to use observation systems setup to study the hierarchical structure of the environment as the initial monitoring data. On the other hand, the more complex the environment, each wave field introduces its information about its internal structure; therefore, the interpretation of the seismic and electromagnetic fields must be conducted separately, without mixing these databases. These results should be used when studying the structure and condition of lunar massive of rocks by deep drilling.

Author details

Olga Hachay[1*] and Oleg Khachay[2]

1 Institute of Geophysics UB RAS, Yekaterinburg, Russian Federation

2 Ural Federal University, Yekaterinburg, Russian Federation

*Address all correspondence to: olgakhachay@yandex.ru

References

[1] Prigogine I, Stengers I. Time, Chaos, and Quantum: To Solve the Paradox of Time. Book House LIBROKOM; 2009. 232 p

[2] Hawking S. From the Big Bang to the Black Holes. A Short History of Time. M. Mir; 1990. Ch. 8

[3] Galkin IN, Shvarev VV. The Structure of the Moon—New in Life, Science, Technology Series "Astronautics, Astronomy". No. 2, Knowledge. 1977. 63 p

[4] Nikolaev AV, Sadovsky MA. New methods of seismic exploration: Development prospects. Herald of the USSR Academy of Sciences. 1982;**1**:57-64

[5] Cosmo chemistry of the moon and planets. (ed.) A. Vinogradov M. editor. Science; 1975. 764p

[6] Kuskov OL. Constitution of the Moon: 3. Composition of middle mantle from seismic data. Physics of the Earth and Planetary Interiors. 1995;**90**:55-74

[7] Kuskov OL. Constitution of the Moon: 4. Composition of the mantle from seismic data. Physics of the Earth and Planetary Interiors. 1997;**102**:239-257

[8] Frondel J. Lunar Mineralogy. J. Willey & Sons; 1975. 177 p

[9] Ruskol E. Origin of the Moon. Moscow: Nauka; 1975. 188 p (in Russian)

[10] Zharkov VN. The Internal Structure of the Earth and Planets. M.: Science; 1983. 413 p

[11] Gusev A, Hanada H, Petrova N. Rotation, Physical Libration, Internal Structure Multiply Moon. Kazan, Kazan Federal University; 2015. 323 p

[12] Sagitov MG. Moon's Gravimetry. M. Science. Main Edition Phys.-Mat. Literature; 1979. 432 p

[13] Vanyan LL et al. The apparent electrical resistance of the moon and its interpretation. Izv. Academy of Sciences of the USSR. "Physics of the Earth". 1973;**11**: 3-12

[14] Khachay OA, Pogrebnoy VN, Khachai OYu, Malosiyeva MT. Similarity of a possible reason for the manifestation of foci-type earthquake foci and dynamic phenomena in impact-dangerous ore mines. Mining Informational Analytical Bulletin. 2018;**4**:159-171. DOI: 10.25018/0236-1493-2018-4-0-159-171

[15] Hachay OA. The study and control of the state of mountain ranges from the perspective of the theory of open dynamic systems. Mining Informational Analytical Bulletin. 2013;**7**:145-151

[16] Hachay OA, Khachay AY, Khachay OY. Chapter 5. Dynamic model of the state of the environment. In: Quadfeul S-A, editor. Fractal Analysis and Chaos in Geosciences. Croatia: InTech; 2012. 174 p

[17] Hachay OA, Khachay AY. The study of the stress-strain state of hierarchical environments. In Proceedings of the Third Tectonophysical Conference at the Institute of Physical Problems RAS; October 8-12, 2012. Moscow: Institute of Physical Problems RAS; 2012. pp. 114-117

[18] Oparin VN, Vostrikov VN, Tapsiev AP, et al. About one kinematic criterion for predicting the limiting state of rock massifs according to mine seismological data. FTPRPI. 2006;**6**:3-10

[19] Hachay OA, Khachay OY. Algorithm for constructing a scenario for the preparation of rock bursts in

rock masses under the influence of explosions according to a seismic catalog. Mining Informational Analytical Bulletin. 2014;**4**:239-246

[20] Chulichkov AI. Mathematical Models of Nonlinear Dynamics. 2003; M. Phizmatlit. 294 p

[21] Hachay OA, Khachay OY, Klimko VK, Shipeev OV. Informative signs of the preparation of high-energy dynamic phenomena according to mine seismological monitoring. Mining Informational Analytical Bulletin. 2015;**4**:155-216

[22] Hachay OA, Khachay OY. Comparison of the features of the synergistic properties of the state of a shock-hazardous rock mass, determined according to seismic and inductive electromagnetic monitoring. Monitoring Science and technology. 2014;**3**:50-55

[23] Hachay OA, Khachay OY. The theory of catastrophes is one of the basic components of the analysis of seismic responses of a mountain range to explosive effects. Fundamental and Applied Issues of Mining Sciences. 2017;**4**(2):175-181

[24] Arnold VI. Catastrophe Theory. 3rd ed. Ext. - M.: Science. Ch. ed. Phys.-Mat. Lit; 1990. 128 p

[25] Hachay OA, Khachay OY, Klimko VK. Dynamic characteristics of slow deformation waves as an massif response to explosive effects. Mining Informational Analytical Bulletin. 2013;**5**:208-214

[26] Hachay OA, Khachay OY, Shipeev OV. Study of the hierarchical structure of the dynamic characteristics of slow deformation waves—Response to explosive effects. Mining Informational Analytical Bulletin. 2013;**5**:215-222

[27] Hachay OA, Khachay OY. The method of assessing and classifying the stability of an array of rocks from the standpoint of the theory of open dynamic systems according to geophysical monitoring. Mining Informational Analytical Bulletin. 2005;**6**:131-142

[28] Hachay OA, Khachay OY, Khachay AY. New methods of geoinformatics for the integration of seismic and gravitational fields in hierarchical environments. Geoinformatics. 2016;**3**:25-29

[29] Hachay OA, Khachay OY, Khachay AY. New methods of geoinformatics monitoring wave fields in hierarchical environments. Geoinformatics. 2015;**3**:45-51

[30] Hachay OA, Khachay AY. Simulation of seismic field propagation in a layered-block elastic medium with hierarchical plastic inclusions. Mining Informational Analytical Bulletin. 2016;**12**:318-326

[31] Hachay OA, Khachay AY. Determination of the surface of anomalously intense inclusion in a hierarchical layered-block medium according to acoustic monitoring data. Mining Informational Analytical Bulletin. 2016;**4**:354-356

[32] Hachay OA, Khachay AY. Determination of the surface of fluid-saturated porous inclusion in a hierarchical layered-block medium according to electromagnetic monitoring. Mining Informational Analytical Bulletin. 2015;**4**:150-154

Section 3

Lunar Surface and Humans

Chapter 5

Lunar Occultation

Abdulrahman Malawi

Abstract

A detailed explanation of the reduction method used to determine the angular diameters of the stars occulted by the dark limb of the Moon is presented.

Keywords: Moon, ephemeris time, occultation, stars, angular diameters

1. Introduction

The disappearance of a star behind the dark limb of the Moon yields very useful information about the star and the lunar surface near that point. This phenomenon is known as lunar occultation. It is in fact part of a very well-known astronomical event—occultation—the disappearance of a nearby celestial object behind another one. The disappearance of the sun behind the Moon, total solar eclipse, the disappearance of planets behind the Moon, or the disappearance of stars behind planets or asteroids are a few examples of this scientifically important phenomenon.

Many scientists have been studying occultation for more than a century (e.g., [1–5]). During the occultation the intensity of the occulted star drops to zero in a very short time. Scientists noticed that the time of the disappearance of the occulted star could be used to determine the diameter of the star. For example, an occulted star with an angular diameter of 0.001 arcsec disappears in 1/50 s, which could easily be measured with photographic recording methods available at that time. Occultation investigations then followed consequently, and a remarkable observation was reported by [6] who used a spinning photographic plate to detect an occultation of Regulus. Diffraction fringes were observed, from which he deduced a diameter of 0.0018 arcsec for the star. This value was in substantial agreement with a later value of 0.0013 arcsec measured by utilizing the intensity interferometer at Narrabri [7]. Scientists in South Africa [8–10] used photographical techniques to make sequence of occultation observations of the red supergiant star Antares, from which they determined the angular diameter of the star. The diameter of μ Geminorum was also determined [11]. The values of the angular diameters of these two stars, which were not affected by the diffraction effects, were found to be 0.040 and 0.023 arcsec, respectively. Ref. [12] also discussed the effect of lunar surface irregularities in the case of stars with large angular diameters. Modern powerful computers and state-of-the-art CCD cameras, along with readily available fast-speed direct digital data recorders, made it very easy for occultation observations to be carried out and for precise models to be constructed.

Lunar occultation observation is the most powerful technique used, to date, for high angular resolution measurements of stars, which made it so important and increases its growth of interest [13–17], adding to the wide range results that can be obtained from analyzing the data. Starting up a lunar occultation observation program is cost-effective (inexpensive) and can be achieved with small- as well as

IntechOpen

large-sized telescopes. Therefore, such program should be considered by medium-sized observatories, even amateur individual observers, making it possible to achieve millisecond resolutions with 1–2 class telescopes fitted with fast photometers [18].

The measurement of stellar diameters is not the only application of this technique and, indeed, may not even be the most fruitful one. Occultation timings are used by the Nautical Almanac offices for the determination of ephemeris time and to improve the theory of the Moon, while the incidental discovery of double stars has been frequent but almost totally unexploited. The event creates a good ground base for an amateur astronomer to do something meaningful in their lives.

2. Theory of occultation and the reduction method

The increased efforts made to time occultation observations, very accurately, resulted in obtaining a detailed information about the limb profile of the Moon. Thus, several observatories established and run observing and analyzing programs. The observed light curves are used to determine the stellar angular diameters, as well as discover double and multiple star systems and map the lunar limb irregularities (see [19] and references therein).

As the Moon precedes between a point on the Earth's surface and the above background stars, it casts down a shadow that moves at a high speed (**Figure 1**).

If the star is observed by a telescope during the shadow passage, the intensity of the starlight would be seen to fluctuate as a result of the movement of the shadow across the aperture of the telescope and ultimately disappears when the lunar shadow becomes total.

The record of these fluctuations constitutes an occultation observation. The intensity pattern F(w) of a monochromatic point source obscured by a straight sharp edge can very well be described in terms of Fresnel integrals ([11]) as

$$F(w) = I_o \frac{1}{2}\left[\left(\frac{1}{2} + C(w)\right)^2 + \left(\frac{1}{2} + S(w)\right)^2\right] \tag{1}$$

where C(w) and S(w) are the Fresnel integrals and they are written as

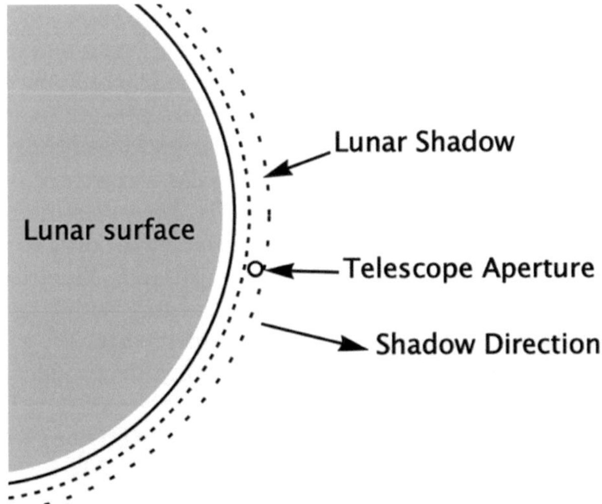

Figure 1.
Diffraction fringes from the lunar limb cross the telescope objective.

$$C(w) = \int_0^w \cos\left(\frac{\pi}{2}t^2\right) dt \quad (2)$$

$$S(w) = \int_0^w \sin\left(\frac{\pi}{2}t^2\right) dt \quad (3)$$

where

$$w = x\left(\frac{2}{\lambda D}\right)$$

Hence the observed light curve is

$$f(x,\lambda) = F(w) \quad (4)$$

where x is the distance in meters from the telescope to the edge of the lunar shadow, λ is a single wavelength, and D is the distance of the Moon from the observer. **Figure 2** shows a monochrome (500 nm) Fresnel diffraction light curve, while **Figure 3** shows the effect of a finite band pass (380–420 nm) on the light curve.

Practically, the diffraction pattern, defined by Eq. (4), is altered by various factors related to the wavelength dependence (detector response, filter band pass, and source dependencies), telescope objective, source size, and the like, as shown in Eq. (5):

$$F_j = E_\circ + I_\circ \iint S(\lambda) f(x,\lambda,\phi)\, D(x)\, \phi(R)\, d\lambda\, dx \quad (5)$$

$S(\lambda)$ comprises the sensitivity function of the detector and the filter, as well as the spectral distribution of the star. It is very important to notice that large telescopes can cause blurring of the fringe patterns, which results in loss of some information, but this is recompensed by the improvement of the S/N ration. Since both finite band width and telescope aperture result in blurring the finer fringes of the light curve, both of them limit the effective spatial resolution in a similar, though not identical, manner. For example, a 2-meter telescope limits resolution to about 0.001 arcsec, while a bandwidth of about 200A (centered at 4000 A) has about the same effect.

The finite disk of the star (**Figure 4**) has the same effect as the telescope aperture, as shown in **Figure 5**. An observer at the center of the Earth sees the Moon moves across stellar background at a rate of about 0.5 arcsec per second, i.e., a fringe passing rate of 0.9 m/s for a star occulted at the leading side of the Moon.

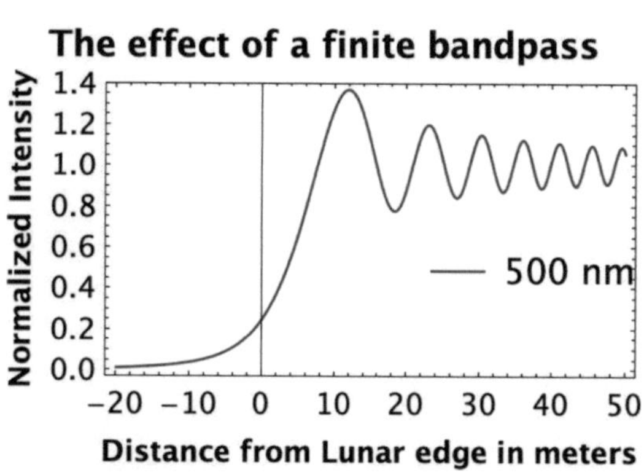

Figure 2.
Monochrome (500 nm) Fresnel diffraction light curve.

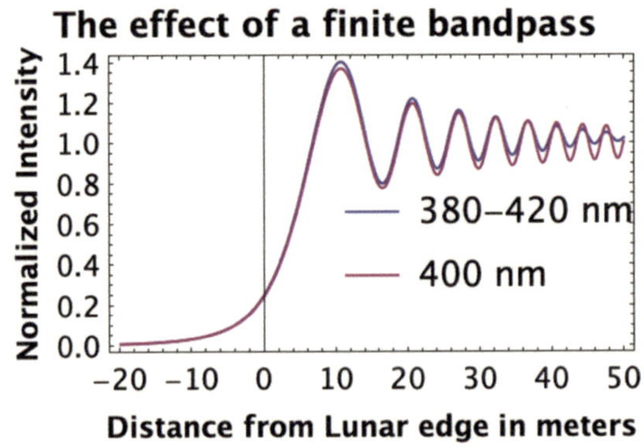

Figure 3.
The effect of a finite bandpass (380-420 nm) on the fine fringes of light curve.

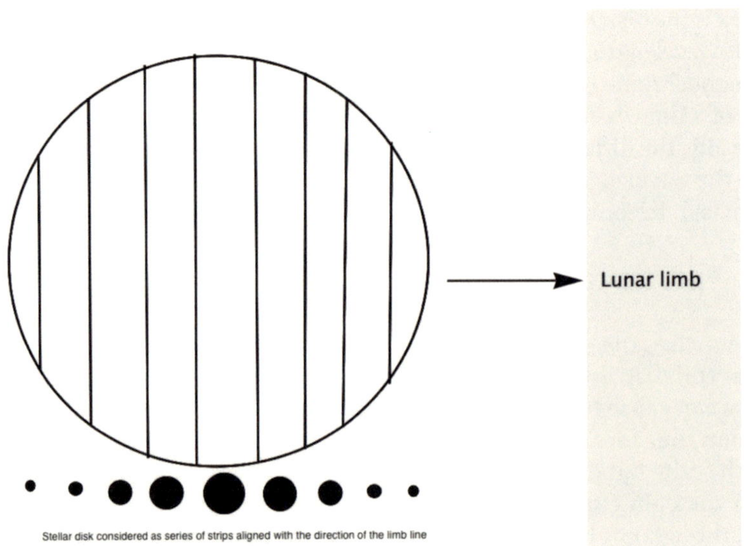

Figure 4.
Stellar disk considered as series of strips aligned with the direction of the limb line.

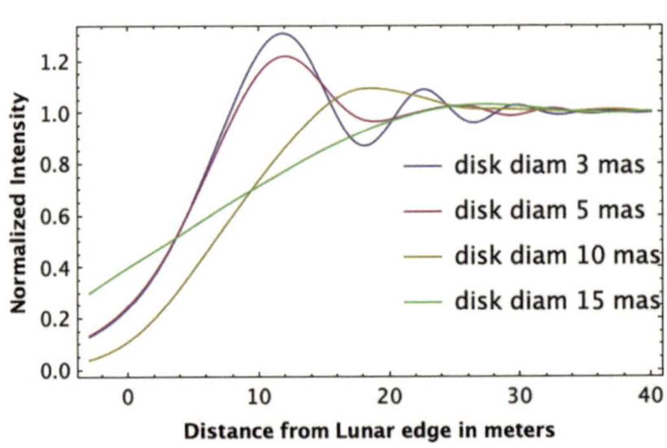

Figure 5.
The effect of uniform circular disk on fringe patterns.

The time scale of the event has been shown to be altered by the angle between the contact point of the occulted star and the motion vector of the Moon ([20]); therefore, scientists proposed that the most appropriate occultation events would be

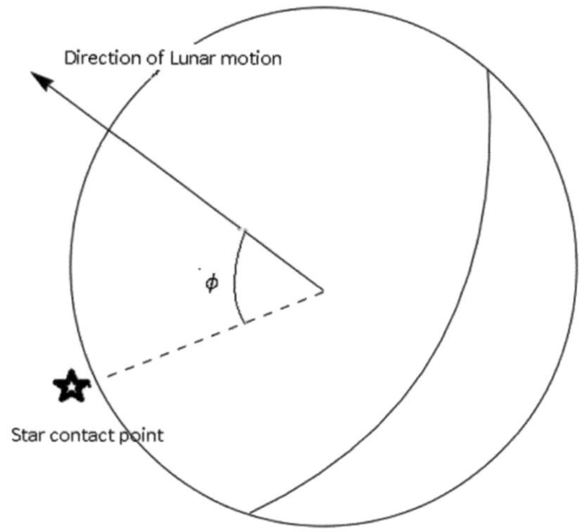

Figure 6.
Occultation time scale is extended by the factor of sec ϕ.

those grazing the edge of the Moon. The effective fringe passage time, as seen from the Earth, is changed by a factor $sec(\phi)$, where ϕ is the angle between the direction of lunar motion and lunar limb at contact point (see **Figure 6**).

The abovementioned effect can improve the quality of an occultation trace. An additional effect on the time scale should also be accounted for. For example, since the Earth rotates in the same direction of the lunar motion, the rate of the lunar motion is decreased due to change in parallax. An equatorial observer, observing when the Moon is on the meridian, near the zenith, will see 50% reduction in the lunar rate. This effect varies as latitude is increased and as the Moon approaches the horizon.

The time scale of fringe passage is also affected by the slope of the lunar surface; therefore, the difference between the obtained parameter of the velocity and the computed velocity is supposed to be caused by the slope of the lunar surface, and one can, in this case, determine its value.

3. Data fitting

The most popular method of analyzing occultation data is the least square method, first introduced by [21]. The Levenberg-Marquardt method of nonlinear fitting subroutine ([22]) fits the model light curve to the observed data of the occulted star. Most of the computed time is taken by Fresnel integrals C(w) and S(w). Therefore, one must use the fastest approximations. Tchebycheff approximations [23] found in [24] is used [25]. It takes the form

$$C(w) = \left(\frac{1}{2} + a(w)\, sin\left(\frac{\pi}{2}w^2\right) - b(w)\, cos\left(\frac{\pi}{2}w^2\right)\right) sign(w) \tag{6}$$

$$S(w) = \left(\frac{1}{2} - a(w)\, con\left(\frac{\pi}{2}w^2\right) - b(w)\, sin\left(\frac{\pi}{2}w^2\right)\right) sign(w) \tag{7}$$

with

$$a(w) = \frac{1 + 0.926|w|}{2 + 1.792|w| + 3.104w^2} \tag{8}$$

$$b(w) = \frac{1}{2 + 40124|w| + 3.492w^2 + 6.670|w|} \quad (9)$$

The program, simultaneously, adjusts five free parameters (the velocity of the shadow A1, the occultation occurrence time A2, the angular diameter of the star A3, the amplitude of the observed light curve at continuum A4, and background level of observed light curve A5) until best fit is obtained. An initial estimate of these parameters must be fed to the computer, adding to the values of another three fixed parameters, representing the observed wavelength, filter bandwidth A6, the Earth-Moon distance A7, and the contact point of the occulted star at the lunar limb A8. The adjustment is carried out by the least square method, mentioned above. The relation between the independent parameters and the values of the Fj is nonlinear; therefore, the program linearizes the equation and iterates until a convincing solution (small χ^2 value) is obtained. The fitting parameters must be provided with expressions for the derivatives of the adjusted parameters with respect to the model function as follows:

$$\frac{dF_j}{dA_1}, \frac{dF_j}{dA_2}, \frac{dF_j}{dA_3}, \frac{dF_j}{dA_4}, \frac{dF_j}{dA_5} \quad (10)$$

When the best fit is found (with minimum value), the program displays the final values of the adjusted parameters, which are supposed to represent the physical states of the occulted star.

4. Estimating the error of the angular diameter measurement caused by scintillation

In order to check our procedure, and to estimate the effects of noise on the calculated parameters, we produce several artificial light curves that represent stars with uniform disks and then modify each curve by adding scintillation noise to it.

The simulated occultation light curves were generated by arbitrarily choosing a portion of each scintillation measurement of 120 data points' length, normalizing it to one, and then adding it to the same length and time resolution of the artificial Fresnel patterns, to give a light curve exactly the same as the light curve resulting from a real occultation. See [25] for more details.

5. Results

Timing and observing lunar occultation events have very important scientific results. For example, grazing occultation observation can be used to gain valuable information about the limb profile of the Moon as well as the occulted star. By combining modern observing equipment and precise time-keeping devices, one can obtain very precise measurements. **Figure** 7 illustrates the event of a single grazing occultation, where eight events were observed. The first disappearance is shown on the left and the last reappearance on the right. The carvery line is the lunar terrain.

Figure 8 (taken from [26]) shows a nearly grazing disappearance occultation observation of the 3.1 mag, A5 spectral-type star 9β Capricorn, SAO 163481, also known as Dabih. This is the main star in a multiple star system. The observation was carried out on 22 Oct. 1993, with a fast photometer, at a sampling rate of 1 mas per data point. Fine details of the fringe pattern of the light curve are perturbed and deformed by scintillation noise, which is the main source of noise,

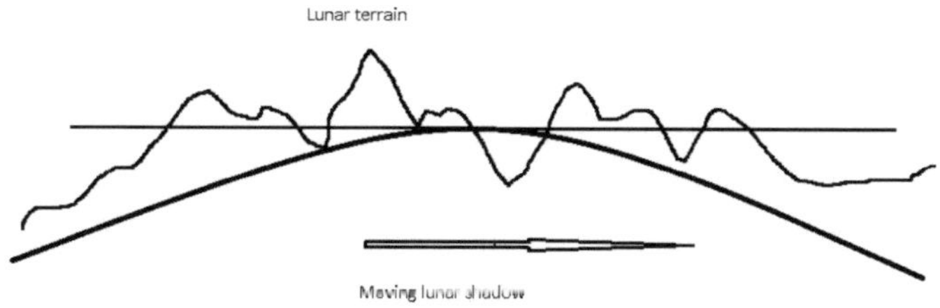

Figure 7.
Illustration of the event of a single grazing lunar occultation.

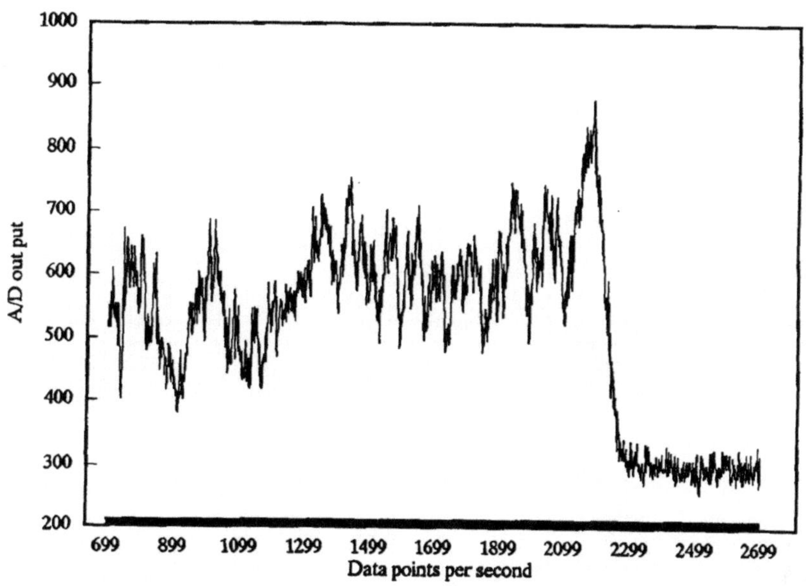

Figure 8.
Light curve of a near grazing lunar occultation of the star 9 β Capricorn, image taken from Malawi et al., 1994.

especially for observations that are carried out at large zenith distance. Other effects, such as irregularities of the lunar limb, and the time delay of the fringe passage caused by nearly grazing occultation can also be seen in the resulting light curve. The best way to minimize the scintillation effects as well as effect of irregularities in the lunar limb is to have multiple observations of the same event whenever possible. If a team of many observers, together, observed the same event, from different locations, perpendicular to the occultation path, they can construct a very accurate profile of the lunar terrain, in the northern and southern pole areas of the Moon.

An example that shows ability of the model-generated curve to fit the observed data is shown in **Figure 9**, where the dots represent the observed data points, whereas the solid line shows the best fit model. The occulted star is 6.69 mag, A7 spectrum (SAO 97647) that was observed on 6 May 1995. The calculated angular diameter is 2.056 ± 0.69 mas. **Table 1** shows the calculated values of the five variable parameters. The estimated angular diagram of the star is 0.0002, "five folds less than the limited angular resolution of the system which is set to be 0.001."

Scintillation effects on the light curve and the smearing of the fine fringe pattern are clearly seen in **Figure 9**. An additional cause of such distortion is the finite band width of the filter used in the observation.

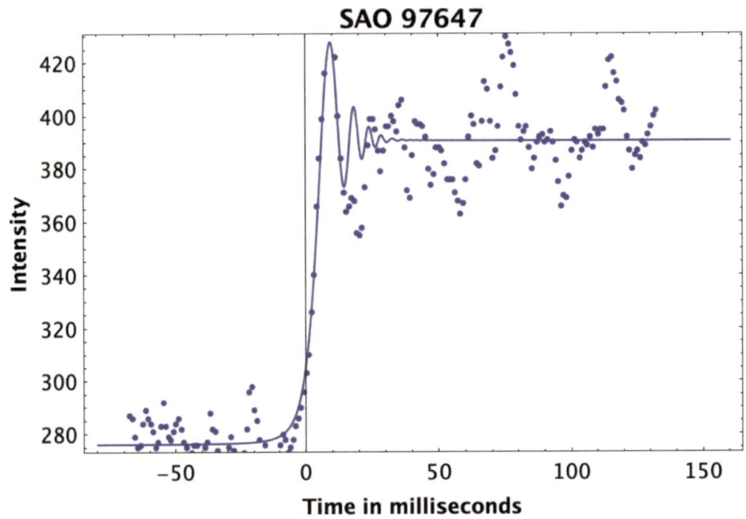

Figure 9.
Observed light curve of the star SAO 97647 (dots) shown along with fitted model (solid line).

Parameter	Estimated	Standard error	Confidence interval
A1	3.58359	0.134282	{3.31876, 3.84841}
A2	0.392329	0.409685	{−0.415628, 1.20029}
A3	2.05561	0.692487	{0.689925, 3.42129}
A4	114.438	1.93875	{110.615, 118.262}
A5	276.119	1.59709	{272.969, 279.268}

Table 1.
The calculated values of the variable parameters.

6. Conclusions

Lunar occultation event offers us a very good chance to study the nature of one or both objects and pave the way for continuing research. For example, grazing occultation observation can be used to gain valuable information about the limb profile of the Moon as well as the occulted star. Observing lunar occultations from different locations is very important, to benefit from the different entering angles of the occulted stars, which helps improve the timing as well as the information about the lunar surface. Although the Japanese Kaguya mission launched in 2007 and the NASA Lunar Reconnaissance Orbiter mission (LRO) in 2009 mapped the lunar surface with an accuracy of 200 down to 5 m, grazing occultations still play an important role in the precision improvement of the lunar terrain features.

Occultation observations, usually, do not require large telescopes. Even telescopes as small as 6 cm refractor or an 11 cm reflector are enough to be used for most occultation events. Since graze paths not often pass over established observatories, amateur astronomers use portable observing equipment and travel to sites along the shadow path limits. The Universal Time Coordinated (UTC) of each event is recorded as accurately as possible, and the observed data is reported to the International Occultation Timing Association (IOTA). By doing so, amateur astronomers can contribute to science. Amateur astronomers can, also, play an important role by discovering unknown double and multiple star systems and/or can investigate them closely through their occultations by the Moon. If timing is made with high resolution, outcome is of high professional interest, to learn more about these systems and to measure stellar diameters. Reductions of grazing occultations also help discover errors in the stellar proper motions as published in the Hipparcos catalog, as well as reveal the rotational error of the Hipparcos reference frame. More details about timing, video or photometric recording, or analyzing lunar occultation observations can be obtained by visiting the web site of the IOTA.

Author details

Abdulrahman Malawi
Kin AbdulAziz University, Jeddah, Saudi Arabia

*Address all correspondence to: amalawi@kau.edu

References

[1] Hough GW. Systematic observations of occultations of stars by the moon. Astronomy Journal. 1902;**24**:191

[2] Robertson J. Reduction of occultations of stars by the moon. Astronomy Journal. 1930;**34**:81

[3] Brouwer D. Occultations of Pleiades stars by the moon. Astronomy Journal. 1932;**42**:18

[4] Brouwer D, Nisoli FE. Occultations of Pleiades stars by the moon. Astronomy Journal. 1935;**44**:95

[5] Andrews LB. Occultations of stars by the moon. Astronomy Journal. 1931;**41**:119

[6] Arnful MA. Sur une méthode pour la mesure des diameters apparents des étoiles. Comptes Rendus de l'Académie des Sciences. 1936;**202**:115

[7] Brown RH. Stellar interferometer at Narrabri observatory. Nature. 1968;**218**:637. Cambridge Astronomy with your personal computer software

[8] Evans DS. The occultation of Antares of 1950 June 27-28. MNRAS. 1951;**111**:64

[9] Cousin AWJ. Photoelectric observations of occultations at the cape observatory. MNRAS. 1953;**113**:776

[10] Evans DS, Heydenry Ch JCR, van Wyk JDN. Observations of occultations of Antares with the Radcliffe reflector. MNRAS. 1953;**113**:781

[11] Evans DS. The angular diameter of Mu Geminorum. MNRAS. 1959;**18**:158

[12] Evans DS. Occultations and lunar mountains. Astronomy Journal. 1955;**60**:60

[13] Richichi A, Chandraschar T, Lisi F, Howell RR, Meyer C, Rabbia Y, et al. Sub-milliarcsecond resolution observations of two carbon stars: TX Piscium and Y Tauri revisited. Astronomy and Astrophysics. 1995;**301**:439

[14] Richichi A, Salinari P, Lisi F. Evidence of pulsation and circumstellar shells in late-type giants obtained by means of lunar occultations. The Astrophysical Journal. 1988;**326**:971

[15] Richichi A, di Giacomo A, Lisi F, Calamai G. Accurate angular diameter and effective temperature of seven late-type stars. Astronomy and Astrophysics. 1992;**265**:535

[16] Bogdanov MB, Cherepashchuk AM. Estimation of brightness distributions over stellar disks from an analysis of infrared observations of lunar occultations - the red giants Sw-Virginis and Fy-Librae. Soviet Astronomy. 1990;**34**(4):393

[17] Ridgway ST. Considerations for the application of the lunar occultation technique. Astronomy Journal. 1977;**82**:511

[18] Richichi A. Interferometry and lunar occultation: Status and prospects of ground-based, milliarcsecond resolution observations. Revista Mexicana de Astronomía y Astrofísica (Serie de Conferencias). 2004;**21**:247-250

[19] Richichi A, Glindemann A. Advances in the interpretation and analysis of lunar occultation light curves. Astronomy & Astrophysics. 2012;**538**:A56

[20] Mac Mahon PA. Stars, Fixed, Determination of the Apparent Diameter of a Fixed Star. MNRAS. 1908;**69**(2):126-127

[21] Nather RE, McCants M, Michael M. Photoelectric measurements of

lunar occultations. IV. Data analysis. Astronomy Journal. 1970;**75**:963

[22] Press WH, Teukolsky SA, Vetterling WN, Flannery BP. Numerical Recipes in Fortran 77: The Art of Scientific Computing. 2nd ed. Cambridge: Cambridge University Press; 1992

[23] Abramowitz M, Stegun A. Handbook of Mathematical Functions. New York: Dover Publ; 1965

[24] Knoechel G, von der Heide K. Statistically rigorous reduction of optical lunar occultation measurements. Astronomy and Astrophysics. 1978;**67**:209

[25] Malawi AA. Developing an optical lunar occultation measurement reduction system for observations at Kaau observatory. Earth, Moon, and Planets. 2013;**110**:175-183

[26] Malawi A, Basurah H, Al-Mleaky Y. Starbright high speed lunar occultation photoelectric photometry. IAPPP Communication. 1994;**57**:4-7

Chapter 6

Solar System Exploration Augmented by In Situ Resource Utilization: Lunar Base Issues

Bryan Palaszewski

Abstract

Creating a presence and an industrial capability on the Moon is essential for the development of humankind. There are many historical study results that have identified and quantified the lunar resources and analyzed the methods of obtaining and employing those resources. The idea of finding, obtaining, and using these materials is called in situ resource utilization (ISRU). The ISRU research and development efforts have led to new ideas in rocket propulsion. Applications in chemical propulsion, nuclear electric propulsion, and many other propulsion systems will be critical in making the initial lunar base and future lunar industries more sustainable and will lead to brilliant futures for humanity.

Keywords: in situ resource utilization, ISRU, lunar base, rocket propulsion, systems analysis, specific impulse, nuclear propulsion

1. Introduction

Human and robotic missions have helped humankind see and understand the many resources of the solar system. The resources have been analyzed, and numerous lunar benefits and industries have been suggested [1–10]. The lunar regolith contains many oxides from which oxygen can be extracted. Water ice in permanently shadowed regions (PSRs) and craters may provide the critical resources for a successful lunar base and lunar cities. The new abilities developed on the Moon can be applied to future human and robotic missions to inner planets, the asteroids, and the outer planets. Mission design studies have shown the great benefits of ISRU in increasing the sample return capability of future planetary missions and vastly extending the reach of exploration. For future large-scale human missions, the possibilities of ISRU for of human exploration and finally settlement offer the best opportunities for sustainability and success.

2. Human exploration options

Since the 1950s, numerous mission studies have identified many effective methods of planetary exploration [1–15]. Robotic exploration has employed the methods of orbital mechanics, systems engineering, and propulsion. Human exploration of the Moon has been conducted, but humans have not yet ventured to Mercury,

Mars, and the outer planets. While future human lunar and Mars missions are in the planning stages, the costs of these missions have prevented their implementation. Extensive mission analyses have identified new strategies for human planetary exploration [16–20]. Cost reductions using advanced propulsion are very critical. In almost every propulsion scenario, ISRU will allow more effective robotic missions and human visits to these planetary targets.

3. The Moon, ISRU, and advanced mission planning

The Moon is the first stepping stone to the rest of the solar system. Since the 1950s, lunar mission planning has yielded many scenarios for exploration, base development, resource mining and use, and industrialization. Many visions of human lunar exploration have been developed, and they all address different possibilities for using in situ resources. A few of the past mission scenarios are summarized here for technology comparisons and insights into new technology infusions.

Many recent studies of the Moon and the use of its resources have been completed [21–25]. While lunar oxygen has been the focus of many of the study teams, water ice in permanently shadowed regions (PSRs) has been analyzed in great detail [26, 27]. Both the oxygen and water ice are critical resources for a more self-sustaining lunar base and a lunar economy. In addition, metals from the oxides in the lunar regolith can provide for construction materials, and lunar regolith can be used for effective radiation shielding from galactic cosmic rays and solar flares.

3.1 Lunar mission scenarios

Large-scale and aggressive lunar base construction was studied by Koelle and his teams [18]. With the advent of the Apollo program, it was deemed reasonable to plan for large lunar operations. His teams at NASA created lunar base construction scenarios using Saturn V class rockets. (**Figures 1** and 2) illustrates the potential cost per person and the number of base personnel [18]. Since the time of its publication, many of the model cost assumptions are no longer valid. However, the example is illustrative of the elements that must be included in future cost estimates. While chemical propulsion was used for the flights from low lunar orbit (LLO) to the surface (called the shuttles), nuclear thermal propulsion (NTP) ferries were used for the round

$$\text{Cost/Man-Year} = \frac{\text{Cost/Ferry Mission}}{\text{10-Man Years/Mission}}$$

where:

$$\text{Cost/Ferry Mission} = \text{(Wt in orbit) (Del. Cost)} + \text{(No. Passengers)}$$

$$\text{(Del. Cost)} + \frac{\text{Ferry Cost}}{\text{Reuses}} + \frac{\text{Shuttle Cost}}{\text{Reuses}}$$

Cost assumptions are as follows:

1. Payload Cost to Earth Orbit = 25, 50, 75 dollars/pound
2. Passenger Cost to Earth Orbit = 100, 200 thousand dollars/man
3. Nuclear Ferry Unit Cost = 25, 50 million dollars
4. Nuclear Ferry Reuses = 10
5. Lunar Shuttle Unit Cost = 10, 15 million dollars
6. Lunar Shuttle Reuses = 10

Figure 1.
Lunar base cost assumption [18].

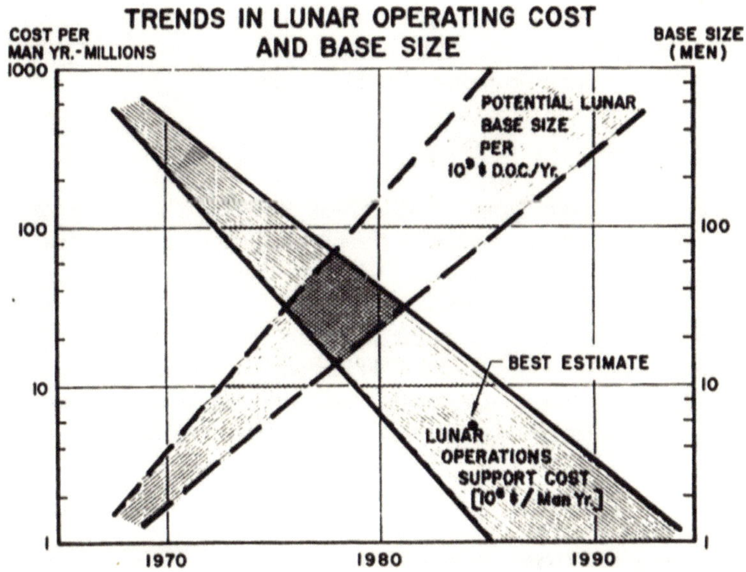

Figure 2.
Lunar base cost and personnel [18].

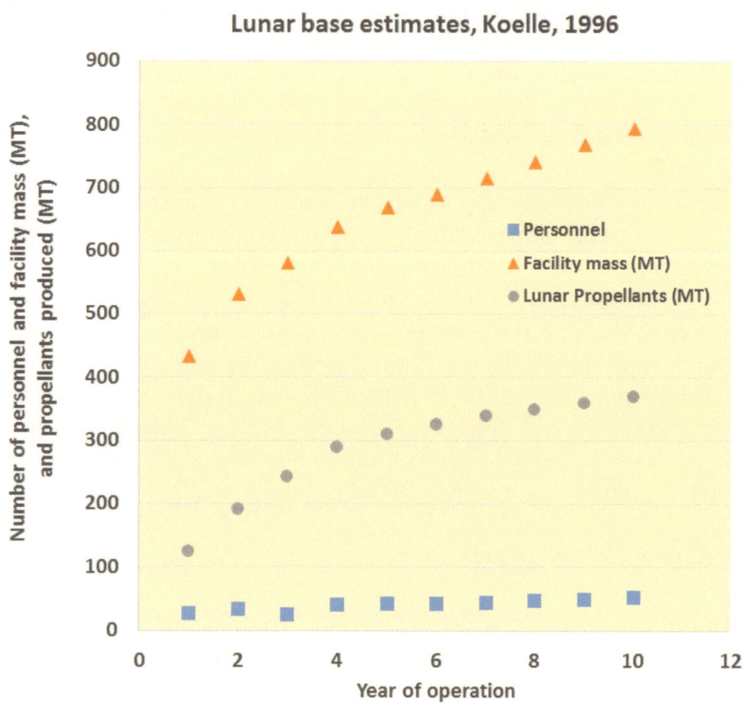

Figure 3.
Lunar base mass, personnel, and propellant produced (derived from [19]).

trips from Earth to LLO. Also, the assumption of ten flights for either the NTP ferry or the chemical propulsion shuttle was included. The NTP ferries carried 20 people with 36.3 metric tons (MT) (80,000 pounds mass (lbm)) of cargo for 6 months of base operations [18].

Later studies by Koelle [19] made more detailed estimates of the lunar base mass and ISRU lunar oxygen production capabilities (**Figure 3**). Over a 10-year period, the lunar base was to be constructed and required approximately 794 MT on the lunar surface. After 10 years, the base would accommodate 52 people and be producing 370 MT of lunar propellants in the tenth year.

3.1.1 Eagle engineering

In 1984, a study was conducted of lunar base construction and the additional accommodation that might be needed at the planned Earth-orbiting space station [20]. Large masses for the lunar base buildup were transported by oxygen/hydrogen orbital transfer vehicle (OTVs) and landers. The OTVs were two-stage vehicles, while the landers were both one-way cargo landers and two-stage human return landers. In this study, 1645 metric tons (MT) of payload was delivered to LLO low lunar orbit. The base would be constructed over a 19-year period. All of the launch vehicles from Earth were space shuttle or space shuttle-derived vehicles.

3.1.2 Spudis

A more modest lunar base scenario has been proposed [26, 27]. In their studies, a more sustainable lunar base was planned. Also, public-private partnerships (PPP) were essential for the success of the lunar base and its ISRU activities. Lunar water ice mining in permanently shadowed regions (PSRs) has been suggested [22]. Mining in the permanently shadowed craters (PSCs) will be challenging. [27] (Commercial Lunar Propellant study, 2018) suggests several solutions to these challenges, which include heliostats to provide lighting in the dark shadowed craters.

3.2 Lunar transportation options

Many techniques have been suggested for reducing the cost of space transportation [28–31]. A recent development is the propulsive landing and reuse of launch vehicle booster stages [28, 29]. While reuse of these launch vehicle stages is a relatively new development, future designs are planned for larger-scale lunar flights [28]. Additional options for lunar exploration and exploitation include a lunar orbital platform or gateway [30, 31]. A gateway may become a central point for propellant storage and distribution to several markets in the Earth-Moon system. These markets included LEO, GEO, LLO, and Earth-Moon libration points [27]. Many study results have identified the potential benefits of these markets, in which the commercial revenue may be many billions of dollars [27].

3.3 Advanced propulsion options

Several advanced propulsion options for lunar base construction and industrialization were investigated. They include nuclear electric propulsion options, lunar base design options, propellant industrialization, and outer planet mining with associated outer planet moon bases. Chemical propulsion and nuclear electric propulsion (NEP) for Earth-Moon orbital transfer vehicles (OTVs) were assessed. Design parameters, vehicle mass scaling equations, and summaries of these analyses are presented.

3.3.1 Chemical propulsion OTV sizing

In sizing the chemical propulsion OTVs, a vehicle mass scaling equation is used [16, 32]:

$M_{dry,stage}\ (kg) = M_{dry,coefficient} \bullet M_p\ (kg).$

where.

$M_{dry,stage}$ = the stage dry mass, including residual propellant (kg).

$M_{dry,coefficient}$ = the B mass coefficient (kg of tank mass/kg of usable propellant mass).

M_p = usable propellant mass (kg).

The chemical propulsion OTVs had a B coefficient of 0.2. The Earth-Moon OTVs were two-stage vehicles (**Table 1**).

3.3.2 NEP OTV sizing

The NEP OTV mass and trip time were estimated based on the power system and the propulsion system design [32]. The following dry mass scaling equation was used [32]:

Mdry,stage (kg) = reactor specific mass (kg/kW) • P (kWe) + 0.05 • Mp (kg) + fixed mass (kg).

The OTV sizing was conducted for a wide range of power levels: 0.5 MWe to 30 MWe. Three nuclear reactor specific masses were used: 10, 20, and 40 kg/kWe (kilograms per kilowatt, electric). The OTV propulsion fixed mass, apart from and in addition to the reactor mass, was 20 MT, and the propellant tankage mass was 5% of the mass of the required propellant.

The Isp and efficiency of the electric propulsion systems were 5000 seconds with thruster efficiencies of 50% for each design. These design points are typical of advanced designs of either magnetoplasmadynamic (MPD) or pulse inductive thrusters (PIT). While hydrogen is suggested for both propulsion system thrusters, the possibilities of the higher Isp option using inert gases (xenon, krypton, etc.) are also viable. The low thrust OTV delta-V value was 16 km/s for the round trip Earth-Moon missions.

Figure 4 shows the propellant masses needed a lunar base scenario; four different propulsion technologies are compared. There is the all-chemical propulsion option and three NEP options with 1, 2, and 5 MWe (megawatts, electric) power levels.

The all-chemical option includes 47 flights of a 35 MT round trip payloads. Each of these OTVs has an initial mass of 155.44 MT. The initial mass of the smaller 6 MT cargo OTV is 26.7 MT. The total propellant loading for the two-stage vehicle is 17.2 MT. To accommodate the 47 human crew flights in each of the NEP options, an 808 MT O2/H2 propellant mass is included.

NEP OTV parameters	Values
Specific impulse (s)	5,000
Engine efficiency (%)	50
Mission payload (MT)	29
Mission delta-V, total (m/s)	16,000

Chemical OTV parameters	Values
Specific impulse (s)	470
Mission payload (MT)	6
Mission delta-V, total (m/s)	5,250
delta-V breakout (m/s):	
Earth departure	3,200
Lunar arrival, LLO	900
Lunar departure, LLO	900
Earth aerobraking	250

Table 1.
NEP and chemical OTV design parameters.

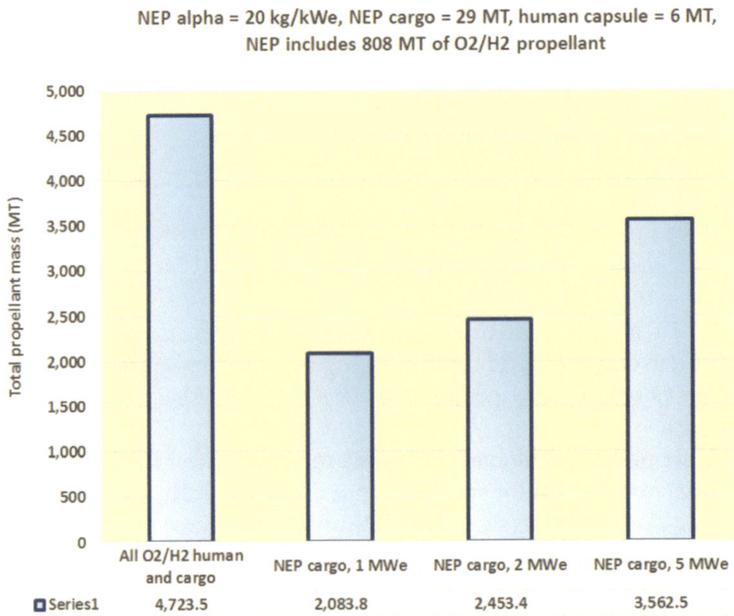

Figure 4.
Chemical propulsion and NEP option comparison (for 1645 MT delivered to LLO).

In all of the NEP options, there are 47 flights of 29 MT payloads. The payloads are carried on the full round trip missions. Once a 29 MT payload is delivered to lunar orbit, it is reasonable to say that a 29 MT payload will be returned to Earth orbit. This payload may be lunar ISRU propellants; lunar landers that may require recycling, updates, or repair; and other finished materials from the Moon.

With advanced nuclear electric propulsion systems, the effectiveness of the lunar base development is enhanced. Using NEP at a reactor alpha of 20 kg/kWe and a 1 MWe power level, over a 19-year assembly period, the propellant mass needed for base transportation can be reduced from 4700 MT to less than 2100 MT. Using NEP at a reactor alpha of 20 kg/kWe and a 1 MWe power level, over a 19-year assembly period, the propellant mass needed for base transportation can be reduced from 4700 MT to less than 2100 MT. Lunar ISRU may allow even further propellant mass reductions. While the NEP trip times are longer for the lower power levels, the overall mass savings is quite significant (**Figure 4**).

3.3.3 Lunar NEP parametric mass analyses

Figures 5–8 provide the initial mass, the propellant mass, and the trip times of the lunar OTVs. The payload mass was 29 MT for the round trip. The range of power levels that were investigated were 0.5–30 MWe. To obtain a balance of trip time and mass savings, a 1–2 MWe OTV is "best" operating point. The broad range power levels and reactor mass scaling parameters in the analyses are presented for additional mission planning purposes.

3.3.4 Lunar lander scenarios

The lander's mission is to deliver lunar propellants or crew or both to the lunar OTV and return to the Moon with cargo from Earth. The round trip delta-V values are provided in **Table 2**. The lander was designed with an oxygen /hydrogen main propulsion system. Lunar lander sizing was conducted for a variety of payload delivery

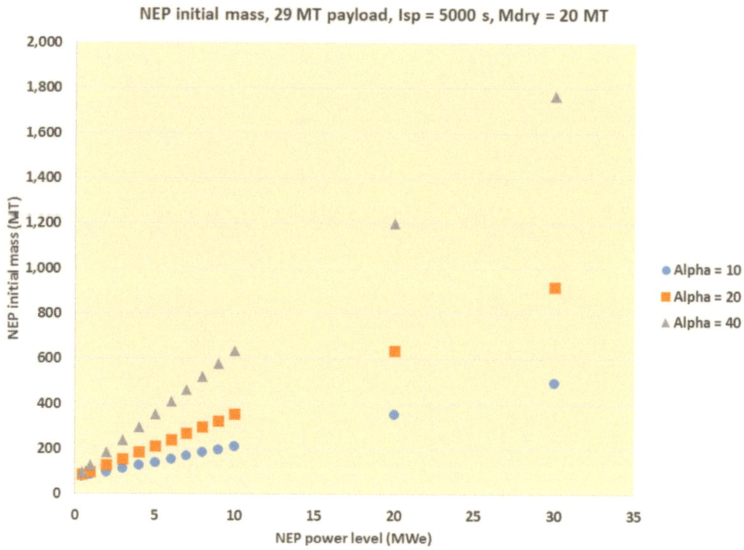

Figure 5.
NEP OTV initial mass versus power level: 0.5–30 MWe.

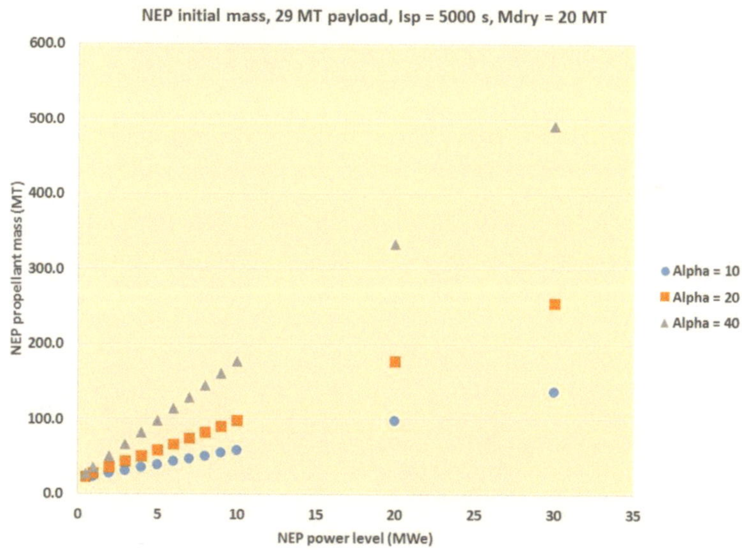

Figure 6.
NEP OTV propellant mass versus power level: 0.5–30 MWe.

missions, rocket engine-specific impulses (Isp), and mission delta-V values. The payload masses were 10, 20, and 50 MT. The rocket engine Isp values ranged from 450 to 480 seconds. Three overall mission delta-V values were selected: 2, 4, and 6 km/s. The 2 km/s delta-V represents a one-way mission from the LLO to the lunar surface. The 6 km/s delta-V represents a lander that can attain near escape velocity conditions about the Moon. Thus, nearly any mission in a wide range of lunar orbits is possible.

Figures 9–11 depict the suggested lunar lander scenarios. For lunar base support, both one-way and two-way lander trips are envisioned. At the outset of the base's construction, there will likely be no significant ISRU propellant production capability; thus a two-way lander is used to assure round trip access from lunar orbit to the base and back to orbit. **Figures 9** and **10** depict the flight profiles for the two-way lander flights. Initial lander flights would begin from orbit and arrive at the surface.

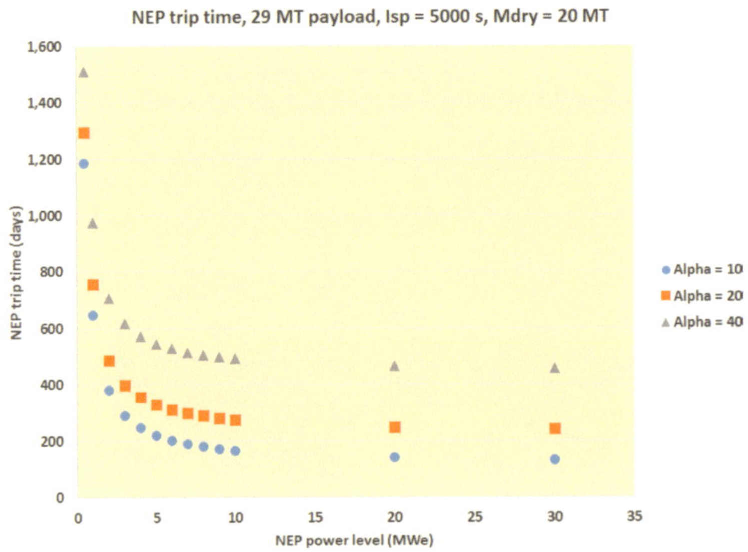

Figure 7.
NEP OTV trip time versus power level: 0.5–30 MWe.

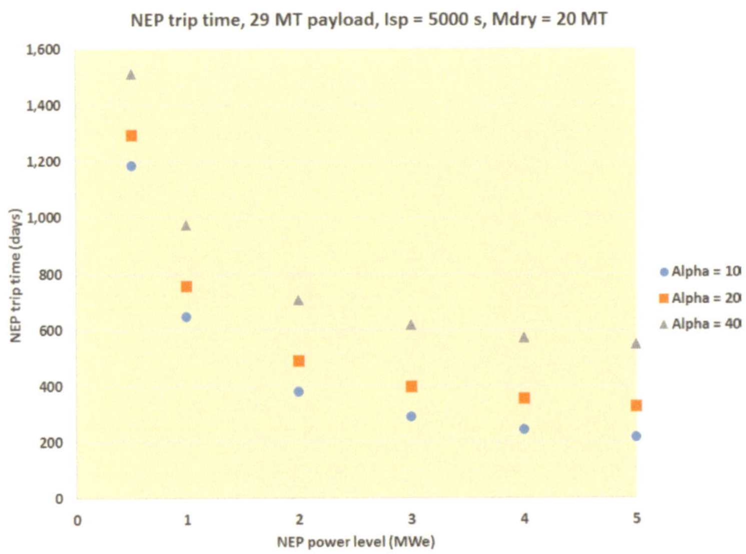

Figure 8.
NEP OTV trip time versus power level: 0.5–5 MWe.

Once a significant ISRU propellant capability is available, a combination of two-way and one-way landers can be used. The one-way lander would have only the propellant capacity to perform a one-way trip, either from orbit to the surface or from the surface to orbit. **Figure 11** illustrates the one-way lander flights. Additionally, a two-way lander can be used to depart from the surface, deliver a payload to orbit, and then return to the surface. A new ISRU-produced propellant load would be available for a subsequent two-way flight.

3.3.5 Lunar lander vehicle masses

The lunar landers are sized with the same mass scaling equations used for the chemical propulsion OTVs. All of the landers were single-stage vehicles. For the lunar

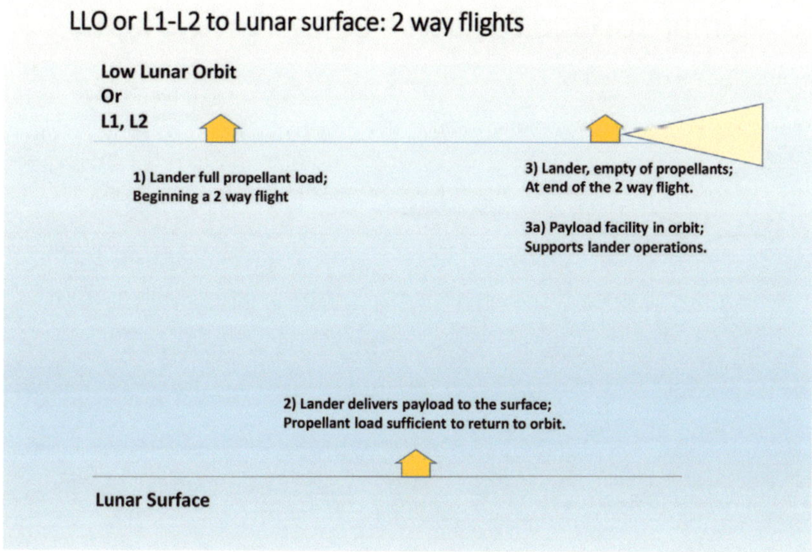

Figure 9.
Two-way lander flight profile with no ISRU.

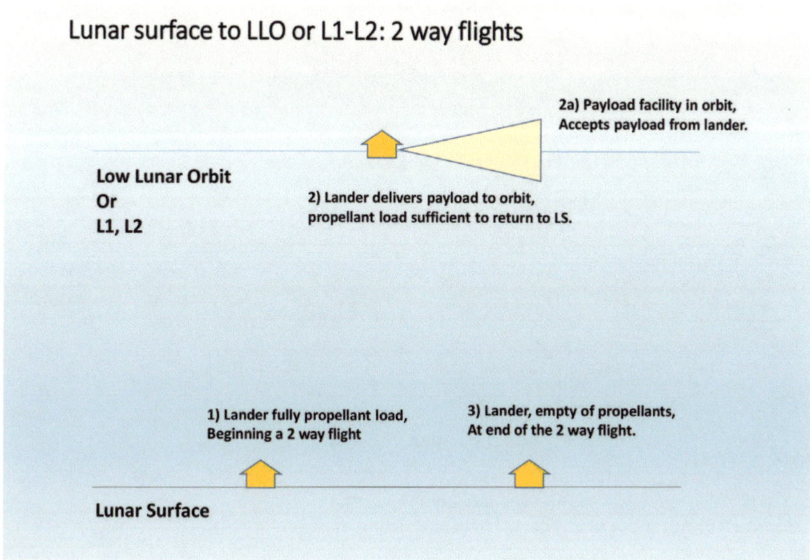

Figure 10.
Two-way lander flight profile with ISRU on the surface.

landers with 2 and 4 km/s delta-V values, the B coefficient was 0.4; in the high delta-V cases for 6 km/s, the B coefficient was 0.2. A 0.2 B coefficient was used as the lander design will not close with a 0.4 B coefficient. As the propellant load is quite high with the 6 km/s lander, and based on historical designs, the 0.2 B coefficient is justified.

Figures 12–14 provide the lunar lander masses for the three delta-V cases: 2, 4, and 6 km/s, respectively. The lander initial mass with a 2 km/s delta-V, a 470-s Isp, and a 10 MT round trip payload mass was 19.7 MT. While such landers will be used

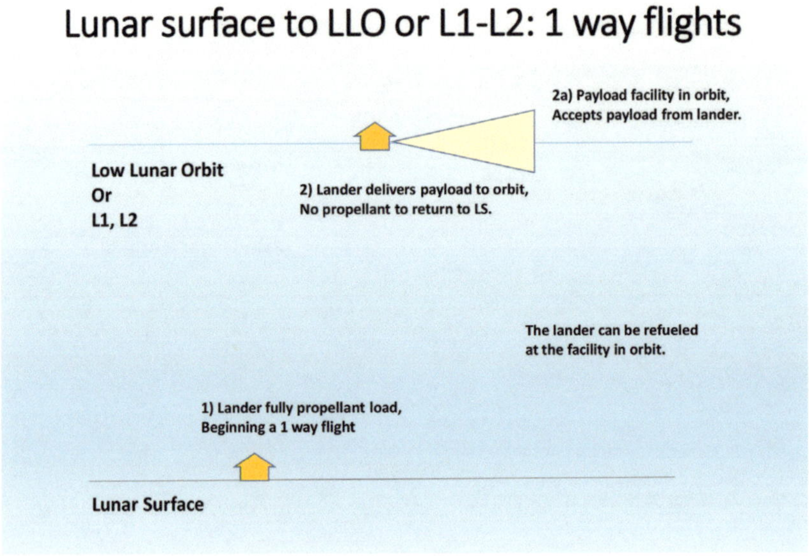

Figure 11.
One-way lander flight profile with ISRU-produced propellants stored on orbit.

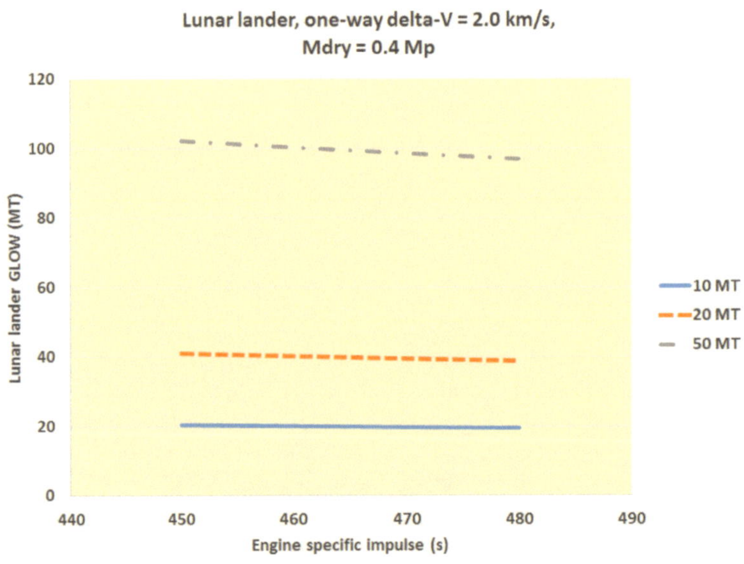

Figure 12.
Lunar lander masses versus specific impulse: 2 km/s delta-V capability.

early in the lunar base construction, these 2 km/s lander cases may also be attractive once lunar oxygen and hydrogen are made available. These smallest delta-V of the landers can be refueled on the Moon to return to lunar orbit. Once there, they can additionally be refueled from on-orbit propellant depots.

The 4 km/s delta-V lander is sized for a round trip with its full payload mass of 10–50 MT. The 4 km/s lander, with a 470-s Isp, and a 10 MT payload, has a mass of 53.2 MT. This delta-V capability offers a propellant load for an abort scenario. If the lander were descending to the Moon, and it were to experience issues during the

Lander parameters	Values
Specific impulse (s)	**450, 460, 470, 480**
Mission payload (MT)	**10, 20, 50**
Mission delta-V (m/s)	**2,000, 4000, 6,000**
delta-V, LS to LLO, one-way	2,000
delta-V, LS to LLO, two-way	4,000
delta-V, LS to libration point, two-way	6,000

Table 2.
Lunar lander mission parameters.

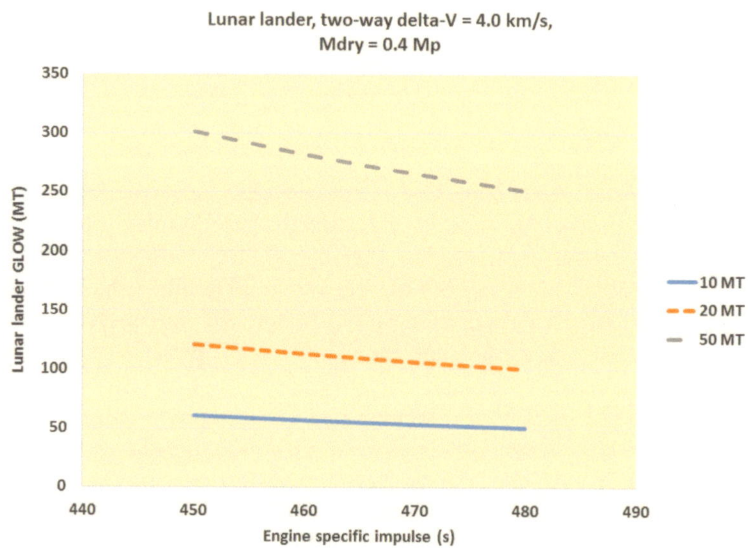

Figure 13.
Lunar lander masses versus specific impulse: 4 km/s delta-V capability.

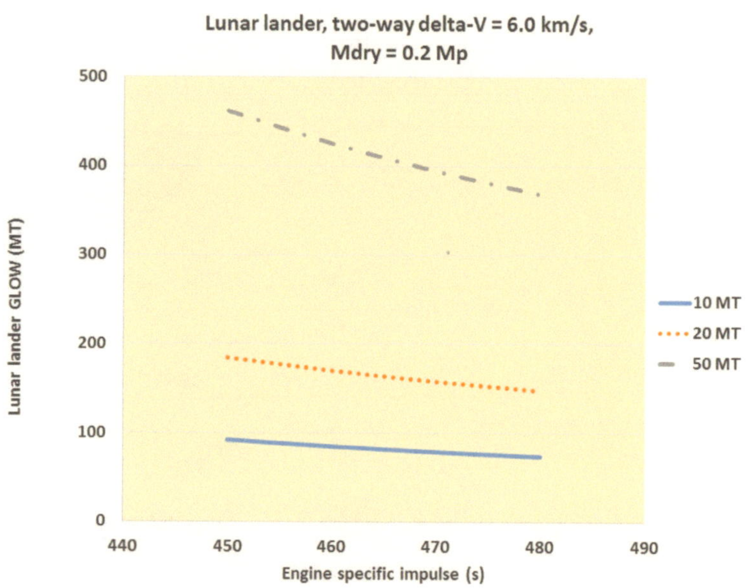

Figure 14.
Lunar lander masses versus specific impulse: 6 km/s delta-V capability.

descent, it would have the full delta-V capability to descend to the surface and then immediately return to orbit without refueling (**Table 2**).

3.3.6 NTP options, with ISRU

Using nuclear thermal propulsion for lunar missions was proposed in the 1960s. Investments and programs to prove the technical feasibility were successful, but these propulsion systems were never flown in space. Since the 1990s, many extensive analyses and experiments have been conducted for nuclear thermal propulsion for lunar and interplanetary missions and demonstrated important payload and trip time benefits [33–35].

A lunar NTP architecture can be refueled with lunar hydrogen, and a specialized design using a liquid oxygen afterburner can increase the thrust level of the lunar NTP shuttle, allowing a shortened 24 hour lunar flight [33].

3.3.7 Lunar base locations

The lunar surface has a wide range of elements available for extraction and use. Lunar water would be most important in sustaining the base. While the Moon has many potential resources available in the regolith, the potential for mining water ice at the lunar poles is strong but challenging. Mining the water will require vehicles that can operate at cryogenic temperatures in the craters. Not only will the robots or other vehicle have to operate in the craters, the light levels will be very low, perhaps requiring operation with light sources fixed at the crater's rim. Onboard power for the robots may have to be provided with nuclear reactors or remotely from a central power recharging station.

Base locations or sites for gathering the water ice must be addressed. The bases in the PSC will be located near the edge of the ice deposit. Locating the base or mining sites at the top (near the crater lip but in the shadow) or the left and right sides of the water ice deposit (and not at the bottom of the crater) were suggested [27]. These sites would provide access to the water ice and remain in the permanently shadowed part of the crater. Potential methods for extracting the water ice are discussed in Refs. [21, 22]. A tent for capturing the water would have a heat source to melt the frozen water ice. A layout for a lunar base is presented in **Figure 15** [36]. The photovoltaic array would be placed outside of the shadowed area, allowing for solar power to support the base and ISRU operations.

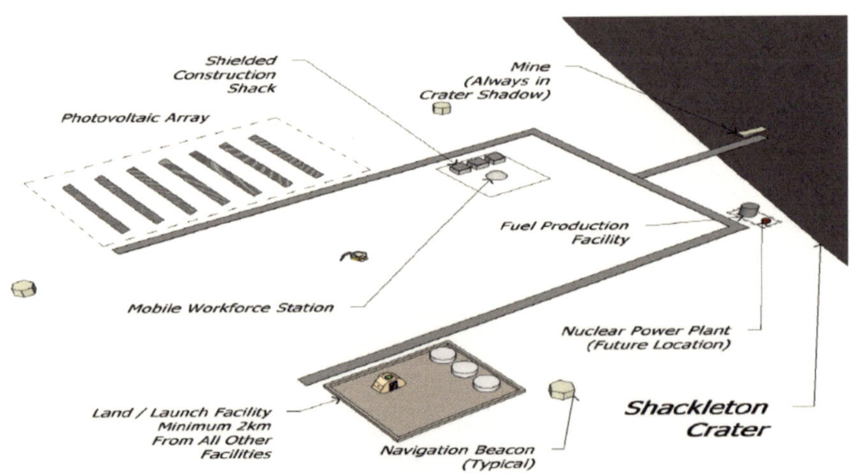

Figure 15.
Lunar base site for mining water ice from a PSC [36].

4. Future lunar base options

4.1 Nuclear underground explosions

Based on recent measurements and simulations of the lunar radiation environment, long-term occupancy of the lunar surface may be detrimental to human beings. In addition to the long-term exposure to natural radiation sources (galactic cosmic rays, solar flares, etc.), there is additional scattered radiation on the lunar surface [28]. Therefore, living and working underground on the Moon may be necessary. Using small of large nuclear devices on the Moon may provide an option for creating large habitable underground spaces. Project Plowshare [37–45] addressed issues with using nuclear devices to complete large-scale civil engineering projects.

Past Earth-based nuclear testing was done underground due to the Nuclear Test Ban Treaty of 1963. The tests often left sizable craters on the surface. When a nuclear device is sufficiently deeply buried, the explosive force can be completely contained underground [39–42]. The blast vaporizes some of the surrounding rocky material which then expands and creates an underground cavity [39–42]. The rocky debris in the cavity undergoes compaction after the explosion, but the initial amount of void space created by the blast just after detonation is distributed in broken rocky debris. Small robotic mining systems would be used for debris removal. Based on historical data, such a space can also be spherical if the blast size is sufficiently small. After the radiation has fallen to acceptable levels, people could potentially create comfortable living spaces.

In Ref. [7], this technique was proposed for not only living spaces but for large-scale ISRU. Ref. [7] illustrates four different ISRU processes using nuclear detonations. There are two chambers: one for the nuclear explosion and one for the reaction product capturing. This processing would essentially chemically reacting oxygen, hydrogen, or other species. The processes range from creating oxygen and metal oxides to producing water and metal carbides.

4.2 Lunar slide lander

In an attempt to reduce the propellant mass needed for lunar landing, the lunar slide lander was conceived. The lander uses friction between a descending tubular spacecraft and a prepared runway of lunar regolith. The operations of the slide lander are in eight phases Ref. [8]:

1. Elliptical orbital descent
2. Perilune pre-landing retro-maneuver
3. Approach to touchdown (begin (vertical) thrust at the end of Phase 3)
4. Touchdown of tail brake
5. Touchdown of side brakes
6. Main drag slide phase with support thrust
7. Main drag slide phase without support thrust
8. Final braking with brief retro-thrust

The slide lander was an attempt to reduce the total propellant load required for lunar landings. While the approach velocity of the lander is over 1.5 km/s, the long slide process may reduce the total delta-V required to 0.2–0.45 km/s; this is a significant delta-V reduction over the 2.0 km/s used for a traditional lunar landing [46]. Using this technique has several drawbacks. The length of the landing strip area is approximately 80 km. Also, the dust from the initial phase of the slide landing may attain an attitude of 1300 of km [8]. Thus, while the landing methods may save landing propellant, the implications of the dust on other orbital operations may be unwieldy.

4.3 Nuclear pulse propulsion

Using nuclear devices for propulsion is another option provided by engineering and physics community [47–50]. The nuclear pulse propulsion (NPP) systems were considered for fast transportation throughout the solar system. Small nuclear devices (physics packages, or PPacks) would be detonated behind a large piloted spacecraft, and the detonation would provide the primary vehicle propulsion. Thousands of nuclear pulses were required for Mars and outer planet missions. The predicted specific impulse for these vehicles is between 1800 and 6000 seconds [47]. The NPP vehicles were considered a logical precursor to the pulsed fusion propulsion systems, noted in many of the atmospheric mining in the outer solar system (AMOSS, [51, 52] studies.

5. Atmospheric mining in the outer solar system

Atmospheric mining of the outer solar system (AMOSS) is one of the options for creating nuclear fusion fuels, such as 3He and deuterium [32, 51, 52]. Uranus' and Neptune's atmospheres would be the primary mining sites. While preliminary estimates of the masses of the mining vehicles have been created [32, 51, 52], supporting OTV and lander vehicles are needed to complete the mining scenarios. Storing the mined gases at automated bases on cryogenic outer planet moons is needed, and lunar base designs for operation in cryogenic environments will be critically important for these outer planet moon base designs.

5.1 Resource capturing studies

Aerospacecraft cruisers have been identified as a "best" solution for atmospheric mining [32, 51, 52]. The main cruiser propellant is atmospheric hydrogen gas, which would be liquefied and used as rocket propellant for the cruise phase and the ascent to orbit. A nuclear gas core rocket is a likely candidate. Deuterium and helium 3 (3He) would be separated from the atmospheric hydrogen, and helium (4He) captured, liquefied, and stored is the primary payload that would be returned to orbit. On each cruiser round trip, a 500 kg payload of deuterium or 3He is captured during the mining time. **Table 3** provides the amount of 3He in the outer planet atmospheres.

	Uranus	Neptune
Amount of 3He in 4He	1.00E-04	1.00E-04
Amount of 4He in atmosphere	0.152	0.19
Amount of 3He in atmosphere	1.52E-05	1.90E-05

Table 3.
Fraction of helium 3 in outer planet atmospheres.

5.2 Vehicle, mission, and propulsion studies

5.2.1 Moon base transportation and mission planning

Several steps are needed to store the nuclear fuels. An aerospacecraft (ASC) must mine the gases from the planet's atmosphere. After mining, the ASC ascends to low orbit and then rendezvous with an orbital transfer vehicle. The OTV and ASC rendezvous at an altitude of at 800 km. After the rendezvous, the OTV accepts the mined cryogenic gases from the ASC, and, the OTV begins a low thrust spiral trajectory to the storage point, an outer planet moon. However, an alternative storage point is an in-space base with artificial gravity; the in-space base would be in orbit about the target Moon. At the Moon, the OTV and outer planet moon lander will rendezvous in high orbit about the outer planet moon. The OTV will deliver the mined fluids to the lander. The lander will refuel the OTV from hydrogen mined on the Moon. The OTV will return to low orbit about Uranus or Neptune to await the next ASC delivery. The lander will return to the Moon with the mined fluids. On the Moon, the lander propulsion system will be refueled with oxygen and hydrogen from the water ice from the Moon. Refs. [23, 42, 43] provide many options for nuclear power and nuclear propulsion to support these mining operations.

6. Observations

Krafft Ehricke envisioned a poly-global civilization, with branches of humanity in many far-flung places in our solar system [1]. His vision was uniquely expressed in Ref. [48]. Here is a short excerpt from that work:

> *Our helionauts, as these men who fly our large interplanetary vehicles call themselves in this era of continuing specialization, have covered the solar system from the sun scorched shores of Mercury to the icy cliffs of the Saturn moon, Titan. They have crossed, and some have died doing so, the vast asteroid belt between Mars and Jupiter and have passed through the heads of comets. Owing to the pioneer spirit, the courage and the knowledge of our helionauts and of those engineers, scientists, and technicians behind them, astrophysicists today work in a solar physics station on Mercury; biologists experiment on Mars, backed by a well-supplied research and supply station on the Mars moon, Phobos; planetologists have landed on Venus; and teams of scientists right now study what have turned out to be the two most fascinating of our solar system, Jupiter and Saturn, from research stations on Callisto and Titan.*

These helionaut flights would be the precursors of human outposts and then colonies all through the solar system. Multiple systems employing planetary ISRU could enable all of these ideas and concepts. The poly-global civilization was considered a natural expansion of the human experience, pioneering new frontiers and using technology in the best interests of all humanity.

7. Concluding remarks

The Moon represents a critical location for the expansion of humanity into the solar system. In an optimistic future, lunar exploration will lead to a base and perhaps extensive lunar industries. The industries include raw material processing, oxygen and other propellant production, nuclear and solar power, and the creation of completely new space vehicles. For protection against radiation, lunar bases may include underground habitats. Using explosive forming of underground cavities may lead to an attractive lunar base or colony. In addition, large-scale mining of lunar raw materials and gas production and capture from underground nuclear processing have been suggested.

With advanced propulsion systems, the effectiveness of the lunar base development is enhanced. Using NEP at a reactor alpha of 20 kg/kWe and a 1 MWe power level, over a 17-year assembly period, the propellant mass needed for base transportation can be reduced from 4700 MT to less than 2100 MT. Lunar ISRU can allow even further propellant mass reductions. With NTP, the payload mass delivered to lunar orbit can be doubled over oxygen/hydrogen chemical propulsion. Further benefits of water mining ISRU can allow refueling of the NTP from lunar hydrogen. Using the option of the liquid oxygen afterburner, the NTP system can allow a 24 hour lunar flight. The added liquid oxygen reduces the NTP Isp but allows a higher thrust level and therefore a shorter flight time. Both the NTP hydrogen and oxygen can be derived from lunar water ice.

Atmospheric mining in the outer solar system can produce nuclear fusion fuels such as 3He which are rare on Earth. In addition, while extracting the small fraction of 3He in the gas giant atmospheres, each day enormous amounts of hydrogen and helium are produced. These amounts can far outstrip the need for propellants to return the mining aerospacecraft (ASC) to orbit. These added propellants may be captured and used for other chemical or nuclear propulsion applications.

Solar system exploration using in situ resource utilization can allow larger and more effective research and sample return missions. Faster missions are possible by using the local planetary resources to return to Earth. Truly impressive interplanetary missions can be within our reach with focused lunar base investments.

Nomenclature

3He	Helium 3
4He	Helium (or helium 4)
AMOSS	atmospheric mining in the outer solar system
ASC	aerospacecraft
CC	closed cycle
delta-V	change in velocity (km/s)
EML1, 2	Earth-Moon libration point 1, 2
GCR	gas core rocket
GTOW	gross takeoff weight
H2	Hydrogen
He	Helium 4
ISRU	in situ resource utilization
Isp	Specific impulse (s)
K	Kelvin
kWe	Kilowatts of electric power
LEO	low earth orbit
LLO	low lunar orbit
LS	lunar surface
MT	metric tons
MWe	Megawatt electric (power level)
NEP	nuclear electric propulsion
NPP	nuclear pulse propulsion
NTP	nuclear thermal propulsion
NTR	nuclear thermal rocket

OC	open cycle
O2	Oxygen
PPB	parts per billion
PSC	permanently shadowed craters
PSR	permanently shadowed regions

Author details

Bryan Palaszewski
NASA John H. Glenn Research Center, Lewis Field, Cleveland, USA

*Address all correspondence to: bryan.a.palaszewski@nasa.gov

References

[1] Ehricke KA. Lunar industrialization and settlement—Birth of poly-global civilization. In: Lunar Bases and Space Activities of the 21st Century Conference; Houston, TX, Lunar and Planetary Institute; 1985. pp. 827-855

[2] Ehricke KA. Harenodynamic cooling - The use of lunar sand as cooling medium. Acta Astronautica. 1984;**11**:319-325

[3] Ehricke KA. Profitability of manufacturing in space in view of lunar industrial development and geo-socio-economic benefits, Manufacturing in space. In: Proceedings of the Winter Annual Meeting, Boston, MA, 13-18, November 1983; American Society of Mechanical Engineers; 1983. pp. 183-198

[4] Ehricke KA. A socio-economic evaluation of the lunar environment and resources. III—Selenospheric economics and cislunar/terrestrial market analysis. In: IAF PAPER 82-235, International Astronautical Federation, International Astronautical Congress, 33rd, Paris, France; 27 September-2 October 1982; p. 25

[5] Ehricke KA. A socio-economic evaluation of the lunar environment and resources. II—Energy for the selenosphere, In: IAF PAPER 79-A-16, International Astronautical Federation, International Astronautical Congress, 30th, Munich, West Germany; 17-22 September 1979; p. 25

[6] Ehricke KA. A socio-economic evaluation of the lunar environment and resources: I. Principles and overall system strategy. Acta Astronautica. 1981;**8**(11-12):1389-1405

[7] Ehricke K. Lunar industries and their value for the human environment on Earth. Acta Astronautica. 1974;**1**:585-622

[8] Ehricke K. Further analyses of the slide lander and of drop delivery systems for improved lunar surface access. Acta Astronautica. 1983;**10**(9):629-650

[9] Ehricke KA. Economy of large launch vehicles including orbital labor cost. AIAA. 1963. p. 277

[10] Ehricke K. A comparison of propellants and working fluids for rocket propulsion. American Rocket Society (ARS) Journal. 1953;**23**(5):287-300

[11] Lawrence DJ et al. Evidence for water ice near Mercury's North Pole from messenger neutron spectrometer measurements. Science. 2013;**339**:292

[12] Manning L. Comparison of several trajectory modes for manned and unmanned missions to Mercury 1980-2000. AIAA. 1967;**67-28**

[13] Marx G. Model study of fuel requirements for fast interplanetary flights using advanced nuclear propulsion systems and refueling at destination [thesis]. German: Technical University of Berlin (West Germany); 1967

[14] Ehricke K. Study of Interplanetary Missions to Mercury Through Saturn with Emphasis on Manned Missions to Venus and Mars 1973/82 Involving Capture. In: Engineering Problems of Manned Interplanetary Exploration; 1963. AIAA. pp. 1963-1514

[15] Ehricke K. Perspective and systems engineering of manned planetary flight, AAS 70-037. In: Space Shuttles and Interplanetary Flight. Vol. 28. Advances in Astronautical Sciences; 1970

[16] Palaszewski B. Metallized propellants for the human exploration of Mars, NASA-Lewis Research Center, NASA TP-3062, presented at the Case For Mars IV Conference, Boulder, CO,

June 4-8 1990. Journal of Propulsion and Power. 1992;**8**(6):1192-1199. Also in the AIAA

[17] Bensky M. Propulsion Requirements for Soft Landing in Extraterrestrial Environments. Rocketdyne, NAS7 121, NASA CR-55088. 1963

[18] Koelle HH, Rutland CH. An example of a reusable earth-lunar transportation system. NASA-TM-X-51436, Paper Presented to the AIAA Summer Meeting, Los Angeles; 01 June 1963

[19] Koelle HH. A Preliminary Concept and Life-Cycle Analysis of a Lunar Settlement. Aerospace Institute, Technical University Berlin, Marchstr.12, D-10587 Berlin, ILR Mitt. 306 (1996); 15 June 1996

[20] Babb G. Impact of Lunar and Planetary Missions on the Space Station. Eagle Engineering, Report Number 84-85D, Contract Number NAS9-17176. November 1984

[21] Li S, Lucey PG, Milliken RE, Hayne PO, Fisher E, Williams J-P, et al. Direct evidence of surface exposed water ice in the lunar polar regions. Proceedings of the National Academy of Science (PNAS). 2018;**115**(36):8907-8912

[22] Stoica A et al. Transformers for Lunar Extreme Environments: Ensuring Long-Term Operations in Regions of Darkness and Low Temperatures. Phase; II. NASA Innovative Advanced Concepts (NIAC), Report Number: HQ-E-DAA-TN58829. 2017

[23] Linne Diane L, Gerald S, Suzuki N, Eisenman D, Hintze P, Araghi K, O'Malley T. Current Activities in the Advanced Exploration Systems ISRU Project. In: Joint Space Resources Roundtable/Planetary & Terrestrial Mining and Sciences Symposium; June 2018

[24] Borowski SK, Ryan SW, Burke LM, McCurdy DR, Fittje JE, Joyner CR. Robust Exploration and Commercial Missions to the Moon Using Nuclear Thermal Rocket Propulsion and In Situ Propellants Derived From Lunar Polar Ice Deposits. NASA/TM—2018-219937, AIAA–2017-5272; November 2018

[25] Connolly JF, Drake B, Joosten BK, Williams N, Polsgove T, Merrill R, et al. The Moon as a Stepping Stone to Human Mars Missions, Paper IAC-18/A3,1,3,x43905. 2018

[26] Spudis P, Lavoie A. Using the Resources of the Moon to Create a Permanent, Cislunar Space Fairing System. AIAA. 2011-7185

[27] Kornuta D. Commercial Lunar Propellant Architecture—A Collaborative Study of Lunar Propellant Production. 2018. Available from: http://www.isruinfo.com/

[28] Gunn JM, Amadieu PJ, Cruzen CA, editors. In: Space Operations: Exploration, Scientific Utilization, and Technology Development: SpaceX Mission Operations. pp. 281-290, Copyright © 2011 by the American Institute of Aeronautics and Astronautics, Inc

[29] Foreman VL, Siddiqi A, De Weck O. Large Satellite Constellation Orbital Debris Impacts: Case Studies of OneWeb and SpaceX Proposals. AIAA 2017-5200, 2017

[30] Smitherman D, Polsgrove T, Rowe J, Simon M. "Safe Haven Configurations for Deep Space Transit Habitats." In: AIAA 2017-5285, AIAA Space 2017 Conference

[31] Polsgrove T, Simon MA, Waggoner J, Smitherman DV, Howard RL, Percy TK. Transit Habitat Design for Mars Exploration. In: AIAA 2018-5143, 2018 AIAA SPACE and Astronautics Forum and Exposition

[32] Palaszewski B. Atmospheric Mining in the Outer Solar System: Orbital

Transfer Vehicles and Outer Planet Moon Base Options. In: AIAA 2016-4889; July 2016

[33] Borowski et al. Vehicle and Mission Design Options for the Human Exploration of Mars/Phobos Using Bimodal NTR and LANTR Propulsion, AIAA 1998-3883; 1998

[34] Bulman MJ et al. High Area Ratio LOX Augmented Nuclear Thermal Rocket (LANTR) Testing. AIAA 2001-3369; 2001

[35] Houts M et al. Nuclear Cryogenic Propulsion Stage. In: Nuclear and Emerging Technologies for Space (2012), 2012

[36] Khan Z, Vranis A, Zavoico A, Freid S, Manners B. Power System Concepts for the Lunar Outpost: A Review of the Power Generation, Energy Storage, Power Management and Distribution (PMAD) System Requirements and Potential Technologies for Development of the Lunar Outpost. In: NASA/TM—2006-214248, DOE/NV/11718-1118; June 2006

[37] Looper M. Radiation environment near the lunar surface CRATER observations and GEANT4 simulations. Space Weather. 2013;**11**:142-152

[38] Cohen GD, Sand FM. Water Resource Applications, Underground Storage of Natural Gas, and Waste Disposal Using Underground Nuclear Explosions. Mathematica Inc Princeton NJ; 31 August 1967. Available from: http://www.dtic.mil/get-tr-doc/pdf?AD=ADA395557

[39] Morrey CB Jr, Pinney E, Stoneham RG, Chambre PL, Lakness RM. Underground Explosion Theory. California Univ Berkeley; 1952. Available from: http://www.dtic.mil/get-tr-doc/pdf?AD=AD0608885

[40] Sachs DC, Swift LM. Underground Explosion Effects. Stanford Research Inst., Menlo Park, CA; 03 March 1958. Available from: http://www.dtic.mil/get-tr-doc/pdf?AD=AD0617181

[41] Brode HL. Nuclear Explosions in Cavities. Santa Monica CA: Rand Corp.; 1965

[42] Brode HL. More Nuclear Explosions in Cavities. Santa Monica: Rand Corp.; 1968

[43] Hoffman GA. Thermoelectric Powerplants Utilizing Contained Nuclear Explosions. Santa Monica, CA: Rand Corp.; 1960

[44] Gruber S. Charged Particle Beam-Plasma Interactions for Thermonuclear Power Generation, Final Report, Case Western Reserve Univ., Cleveland Ohio, Dept. of Electrical Engineering and Applied Physics; 1974

[45] Wong HV, Kotschenreuther MT, Breizman BN, Van Dam JW, Hazeltine Richard D. Assessment of Compact Low Neutron Fusion Reactor Concepts, Final Rept. 15 Jun 1999-30 Nov 2000, Texas Univ. At Austin Inst. For Fusion Studies, 16 February 2000. Available from: http://www.dtic.mil/get-tr-doc/pdf?AD=ADA387427

[46] Palaszewski B. Lunar Missions Using Advanced Chemical Propulsion: System Design Issues. In: NASA-Lewis Research Center, NASA TP-3065, AIAA 90-2341, presented at the 26th AIAA/ASME/SAE Joint Propulsion Conference, Orlando, FL, July, 1990, Also in AIAA Journal of Spacecraft and Rockets. Vol. 31, No. 3, May-June 1994. pp. 458-465

[47] Interplanetary maneuvers in manned helionautical missions. (Ehricke, 1965). AIAA. 1965-695

[48] Ehricke KA. Solar Transportation. In: Presented to the 4th Goddard Memorial Symposium, Am. Astronautical Soc., Washington, D. C., 15-16 March. 1966

[49] A Grand Vision of Man's Role in Colonizing the Universe by Oyang Teng, LaRouche Youth Movement (Book review)—Marsha Freeman, Krafft Ehricke's Moon: The Extraterrestrial Imperative, Technology Editor of Executive Intelligence Review, 2009. Available from: http://www.21stcenturysciencetech.com/Articles_2009/Summer-2009/Extraterrestrial_Imperative.pdf

[50] Schmidt GR, Bonometti JA, Irvine CA. Project orion and future prospects for nuclear pulse propulsion. Journal of Propulsion and Power. 2002;**18**(3)

[51] Palaszewski B. Atmospheric mining in the outer solar system: Resource capturing, storage, and utilization. AIAA. 2012;**2012-3742**

[52] Palaszewski B. Atmospheric Mining in the Outer Solar System: Issues and Challenges for Mining Vehicle Propulsion. AIAA. 2011-6041; August 2011

Chapter 7

Human Health in the Lunar Environment

Robert J. Reynolds

Abstract

The lunar environment contains many hazards to human health, some common to extraterrestrial locations, some unique to the Moon. Exposures of particular concern are hypobaric environments, hypogravity, space radiation, and lunar dust. This chapter provides a brief overview of these exposures, as they represent the gravest threats to human health in the lunar environment (i.e., they may affect mortality rates) and then reviews the published studies of mortality of the original twenty-four lunar astronauts who visited the Moon between 1969 and 1972. The chapter closes with a reexamination of lunar astronaut mortality using updated data, including detailed discussion of the interpretation of the results.

Keywords: astronauts, mortality, cardiovascular disease, cancer, Moon

1. Introduction

Though seemingly serene, the lunar environment presents a number of hazards for human health, both acute and chronic. Many of these hazards are common to low Earth orbit (LEO) or deep space, while others are unique to the Moon. As several national space agencies have declared their intentions for further manned lunar exploration and eventual colonization, it is more important than ever to understand the health risks involved.

This chapter provides an overview of the major hazards associated with space exploration in general and examines how those hazards may differ in the lunar environment. The number of health risks inherent in space exploration is staggering, necessitating that this overview be limited in breadth, covering only those risks that are the most serious: hypobaric syndromes, cardiovascular disease, respiratory disease, and cancers. A thorough discussion of any one of these could easily fill a chapter. For this reason, the review is also limited in depth, aiming to provide adequate context for more detailed studies of astronaut mortality related to these risks.

Next, the chapter reviews the research to date on the mortality outcomes of the original twenty-four lunar astronauts and presents the results of an updated investigation of mortality among lunar astronauts, expanding upon the findings from earlier studies.

2. Environmental hazards and health outcomes in the lunar environment

2.1 Hypobaric environments

Lunar atmospheric pressure has been estimated as 3×10^{-15} bar (3×10^{-12} kPa) at night [1]. This presents hypobaric health risks, requiring astronauts to wear

IntechOpen

space suits or be within the confines of pressurized habitats at all times. The lack of atmosphere on the Moon means that lunar astronauts face the constant threat of decompression sickness when moving between environments of differing pressure, and of ebullism if exposed to the lunar surface without a spacesuit [2, 3].

The danger of decompression sickness is highest for astronauts leaving a lunar habitation or spacecraft to perform an extra-vehicular activity (EVA) on the lunar surface. Spacecraft are typically pressurized to sea-level on Earth, about 101 kPa, whereas US spacesuits are pressurized to only approximately 30 kPa. Astronauts suiting up and exiting the spacecraft or habitation to the lunar surface could develop nitrogen bubbles in the bloodstream or in tissues as dissolved nitrogen condenses under the reduced pressure [2].

Ebullism occurs when bodily fluids vaporize under extremely low atmospheric pressure (or in vacuum). During this process the body will swell as water vapor forms in soft tissues and venous blood. Ebullism is but one effect of exposure to very low pressure; others include cessation of circulation, loss of consciousness, paralysis, and eventual death [3].

2.2 Hypogravity

Life on Earth exists in an environment of constant gravitational acceleration. This ubiquitous force has shaped the evolution of plants and animals on the planet, humans included. The human skeletal structure, cardiovascular system, vestibular system and other physiological characteristics have evolved to function best with a constant downward pull of 9.81 m/s^2 (1 Earth-gravity or 1 G). Astronauts traveling to the Moon will instead encounter lunar gravity, which is approximately 1/6 G. They will also spend time both in LEO and translunar space, which are 0 G or *microgravity* environments. Environments with less than 1 G of gravity may collectively be referred to as *hypogravity* environments.

Hypogravity is directly or indirectly responsible for a number of physiological changes during spaceflight. One of the most important is cardiovascular deconditioning. Cardiovascular deconditioning is the constellation of symptoms that result from weakening of the heart as a result of prolonged periods of decreased cardiac workload. Included among these symptoms are temporary and less serious symptoms such as orthostatic intolerance and reduced aerobic capacity, as well as the potentially more serious outcome of cardiac arrhythmia [4, 5].

Cardiovascular deconditioning during space travel is a result of the redistribution of bodily fluids that occurs under microgravity. Of particular concern is the redistribution of blood throughout the circulatory system. On Earth, the pull of gravity pools blood in the legs and feet, whereas in microgravity the cardiovascular system is easily able to create a more uniform distribution of blood between the upper and lower parts of the body. Once this distribution has been achieved (within hours of entering 0 G), the body perceives the increased volume of blood in the upper half of the body as hypervolemia, and reduces the total volume of water in the body to compensate. The net effect of this is a reduced workload on the heart, which can eventually lead to cardiac atrophy [4, 5].

Short-duration arrhythmias have been recorded on several occasions during spaceflight, but it is unclear if they are the result of underlying, preexisting conditions or are the result of changes induced by spaceflight [6, 7]. One potential explanation for these arrhythmias is the loss of potassium with the reduction of water in the body when adjusting to 0 G [8]. At the time of press, NASA still considers this an open question [7].

Since fluid shifts under reduced gravity are responsible for the cardiovascular changes observed in astronauts after extended stays in LEO, it is logical to assume that the Moon's reduced gravity would lead to a similar, but perhaps lesser, deconditioning. A recent review examined the evidence regarding cardiopulmonary and other outcomes in simulated hypogravity, including simulated lunar gravity. The review found that several physiological measures of cardio-pulmonary efficiency are improved with decreasing gravity levels, and that cardiac stoke volume increases with decreasing gravity, similar to 0 G conditions [9]. However, it is still unknown the extent to which long term exposure to lunar gravity would attenuate the cardiac deconditioning seen with stays in microgravity.

2.3 Space radiation

Just as life on Earth has been shaped by the Earth's gravitational pull, it has also been influenced by the Earth's radiation environment. The types of and amounts of radiation typically found on Earth differ from those in space. In general, the amount of background radiation on Earth is lower than that in space, particularly for cosmic radiation. As a consequence, life on Earth has evolved in an environment comparatively devoid of ionizing radiation, and thus is not generally resistant to it.

Ionizing radiation in outer space is primarily particulate: protons, electrons, and heavy atomic nuclei. Fast-moving ions may come from deep space (galactic cosmic rays, or GCR) or may be ejected from the Sun during a solar particle event (SPE). Protons are also encountered as particles ejected from the Sun as part of an SPE [10, 11]. Since the Moon has very little atmosphere and a weak magnetic field, the lunar surface is unprotected from cosmic radiation and thus under constant bombardment from it [12].

Exposure to ionizing radiation can have effects that occur either relatively soon after the dose (within 24 hours to several weeks) or relatively late after the dose (months or even years later). The short-term effects are called acute effects, while those that occur later are termed late effects. For astronauts living and working on the Moon, there is considerable risk for both acute and late effects.

2.3.1 Acute radiation syndrome

Acute Radiation Syndrome (ARS) is the collection of health effects that occur after a rapid, whole-body dose of ionizing radiation. Typical effects of ARS include retching and vomiting, which has been observed in animal models of SPE irradiation at doses as low as 0.5 Gy. Hematological changes and immune system suppression have been observed in various irradiated animals, as has damage to lung tissues, impaired cardiac function, and skin damage. Finally, fatigue has been observed as a common outcome of acute radiation exposure, both in cancer patients receiving radiotherapy and in animal models of SPE radiation effects [10].

Just as for astronauts in LEO or deep space, the risk of ARS for astronauts in the lunar environment is greatest for astronauts who may be on EVA during an SPE, as they will not have the benefit of spacecraft or lunar habitat shielding [12]. Though the most likely risks attached to SPEs are acute effects, a SPE of sufficient size could prove lethal to lunar astronauts, and such events have been observed in the recent past [12]. Even when SPEs are not of sufficient size to be outright lethal, doses absorbed during an SPE could still result in an ARS episode, and would contribute to the risk of late effects, as detailed below.

2.3.2 Carcinogenesis and heart disease

The primary late-effect health concern from space radiation is radiation-induced carcinogenesis. Leukemias, breast, lung, and gastrointestinal tumors have all been observed after exposure to ionizing radiation on Earth, and thus it is speculated that space radiation could yield similar problems, perhaps at higher rates [11, 13]. While there have been no reports of increased incidence of cancer among astronauts to date, this is likely due to insufficient radiation doses stemming from successful radiation control strategies as practiced by NASA, and relatively brief journeys in LEO or to the Moon [14, 15].

Though epidemiologists continue to surveil the astronaut population for cancer mortality, a growing concern is the degenerative tissue effects that space radiation may have on the cardiovascular system, as accumulated damage to the vasculature may induce heart disease [10, 16]. Recent research has failed to find any increased incidence of heart disease among astronauts, even when controlling for known cardiovascular risk factors. Instead, astronauts were found to have a much lower incidence of cardiovascular disease in comparison to both the US general population and a specially matched control population [17]. Furthermore, there was no difference in risk between astronauts who had spaceflight experience in comparison to those who did not.

2.4 Lunar dust

The surface of the Moon is covered in a meters-thick layer of loose rock and fine dust known as *lunar regolith*. This substance is the result of meteorite collisions with the Moon breaking down exposed lunar bedrock over the course of billions of years [18]. The mean size of regolith dust particles from the Apollo and Luna samples are between 60 and 80 μm, though ultra-fine particles (those with sizes measured in nanometers) have been documented as well [18, 19]. It is estimated that approximately 10% of the very fine particulate component (<10 μm) is in the respirable range [20].

Respirable particles of lunar regolith have unique properties which make them potentially more toxic than dusts found on Earth. Much of the fine particle regolith material is created by localized vaporization and fracturing when micrometeorites impact the lunar surface and the existing regolith material. This creates particles that are jagged and sharp, as well as highly chemically reactive. Without substantial atmosphere and weather to grind down and react with the particles, they remain jagged and chemically reactive [20–22]. Finally, regolith particles of respirable size are near-ubiquitous with nanophase iron (np-Fe^0) spheres (super-fine iron dust with particles measured on a nanometer scale) [21]. These physical and chemical properties of lunar regolith make it potentially harmful for human skin, eyes, and, most importantly, airways.

Lunar regolith poses a respiratory health risk in much the same way that respirable dust on Earth does. The inhalation of respirable particles (<10 μm) into the airway can irritate the airway linings and bronchi, and dust may lodge in the lungs themselves, promoting fibrous growths. This effect may be exacerbated by hypogravity, as simulation studies using animals have shown differential patterns of particle deposition under reduced gravity conditions [23–25]. Because of its jagged shape, lunar dust adheres to spacesuits, and, during the Apollo lunar missions, became airborne inside the lunar and command modules after lunar surface excursions.

Exposure to silica on Earth is known to be associated with pneumoconiosis, increased risk of lung infection, chronic obstructive pulmonary disease, cancers of the airway, certain autoimmune disorders, and chronic renal disease [26]. While it is not known for sure, it is reasonable to assume that lunar dust would also be associated with these conditions, since, as mentioned above, its physical and chemical properties make it more toxic than silica dust on Earth.

Safe human levels of exposure to lunar regolith have been estimated through animal studies [27, 28]. These studies have estimated the same exposure level to be in the range of 0.5–1.0 mg/m^3 for a 6-month lunar surface deployment. Still another study estimated a human no adverse effect level (NOAEL) for intermittent exposure on a 6-month mission as 0.4 mg/m^3 [29]. For comparison, the US Occupational Safety and Health Administration (OSHA) sets a time-weighted average permissible exposure limit (PEL) for respirable fused silica dust of 5 mg/m^3 in an 8-hour shift of work. Respirable fused silica dust currently has no threshold limit value, but OSHA recommends that airborne concentrations of respirable particles be kept below 3 mg/m^3 for insoluble particles of low toxicity [30].

2.5 Other risks

As part of its Human Research Program (HRP), NASA maintains a list of thirty-four health risks associated with deep space exploration [31]. Although they are not organized by cause, at least half of these risks are related to the environmental conditions reviewed here: vacuum, hypogravity, cosmic radiation, and celestial dust (including lunar dust). Many of the rest of the risks are comprised by categories of occupational hazards such as inadequate design and engineering, psychological and performance issues, inadequate training, and the physical hazards associated with EVA. Interested readers are encouraged to investigate the list published by NASA and learn more about these risks.

3. Mortality risk for lunar astronauts

Though humans have spent relatively little time on the lunar surface, health risks associated with the lunar environment are myriad. It is therefore possible that lunar astronauts have faced premature mortality risk because of their time in orbit around the Moon or on its surface. This section first reviews the existing literature on lunar astronaut mortality, and then updates those findings with a new original analysis of the data.

3.1 Prior studies of lunar astronaut mortality

Studies of lunar astronaut mortality have focused on searching for evidence of increased mortality due to either cancer or cardiovascular disease, either of which the literature suggests may be elevated with sufficient exposure to ionizing radiation in outer space.

The first study to isolate lunar astronauts in a mortality analysis attempted to examine if a larger proportion of lunar astronauts had died of cardiovascular disease than might be expected given a general population comparison cohort or in comparison to non-lunar astronauts (including members of pre-NASA experimental flight programs administered by the US Air Force) [32]. To do this however, the authors used a cross-sectional statistical method inappropriate to the data [33, 34]. This approach led the researchers to conclude, erroneously, that lunar astronauts had indeed experienced an excess proportion of deaths due to cardiovascular disease in comparison to the general population, as well as in comparison to astronauts who never went to space and astronauts who only flew on missions to low Earth orbit. A reanalysis of the data (using methods appropriate to longitudinal cohort studies) told a different story: lunar astronauts were at reduced risk of mortality from cardiovascular disease in comparison to the general population, and showed no differences in risk in comparison to other groups of astronauts [34]. This conclusion was further supported by a study of cardiovascular disease incidence which, as described above, found no increase in the incidence of cardiovascular disease for any astronauts [17].

A recent study of cosmic radiation exposure and mortality risk analyzed data from all 73 astronauts selected before 1970 [15]. The study used recorded radiation doses as the exposure measurement, and looked at several cancer and CVD outcomes. Results indicated significant reductions in mortality risk from all cardiovascular disease, ischemic heart disease, all cancers, and all-cause mortality, even though they found a greater than five-fold increase in risk of death by external causes. Though this study was not exclusively of lunar astronauts, lunar astronauts did make up one-third of the study group. These results agree with prior studies, and are further suggestive of a lack of excess mortality risk in the lunar cohort from cancer, CVD or any other natural cause.

3.2 Updated analysis of lunar astronaut mortality

This analysis compares the mortality rate of lunar astronauts to that of all other NASA astronauts for several cause-of-death categories, and with follow-up through 31 October 2018.

3.2.1 Data

Data for this analysis were those of all male NASA astronauts selected between 1959 and 2013. As there have been no female lunar astronauts, I limited the dataset to males. I counted follow-up time for non-lunar astronauts as the date of selection until either death or 31 October 2018, whichever came first. To avoid immortal time bias, follow-up time for lunar astronauts was from the date of the first lunar mission until either death or 31 October 2018 [35]. A total of 314 non-lunar astronauts contributed 7534 person-years of observation time and 50 deaths, while the 24 lunar astronauts contributed 1002 person-years of observation and 12 deaths.

3.2.2 Causal model

Figure 1 displays a causal diagram for astronaut mortality in the form of a directed acyclic graph (DAG). The diagram shows the theorized causal relationship between several measured factors, unmeasured factors, and the outcome, mortality. In this DAG, Year has a direct effect on other potential explanatory variables as well as a direct effect on mortality rate. This makes Year a confounder of the relationship between the exposure of interest – being a lunar astronaut – and mortality rate. This in turn means that the paths leading from Year to Mortality Rate are biasing paths as well. Controlling for Year removes this confounding, so I included Year in all models. Similarly, Age at Selection confounds the relationship between being a lunar astronaut and mortality through its causal effect on being a lunar astronaut and on current age. I therefore adjusted for current age in all models as well.

The circle for "other factors" represents any of an unlimited set of potential confounding variables that were not measured in the data. This set of unnamed factors are also assumed to be influenced by year of selection, and themselves have a causal effect on becoming a lunar astronaut as well as mortality rate. This makes these other factors confounders as well. However, as they are unmeasured, there is no way to control for any residual confounding that these factors may introduce.

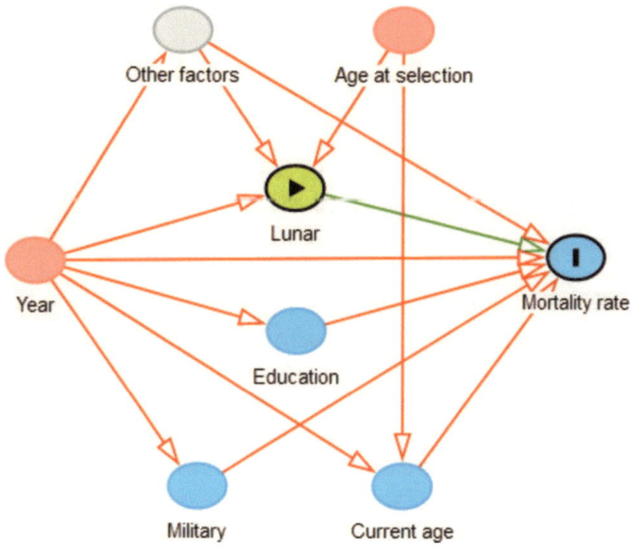

Figure 1.
Causal diagram for lunar astronaut mortality.

3.2.3 Statistical methods

To estimate if lunar astronauts have experienced higher age-specific mortality risk since their lunar missions, I modeled their mortality due to various causes of death. To do this I fit a series of Poisson regression models to discreet time-interval data. These data break the total follow-up period for each individual into smaller intervals with both fixed and time-varying covariates, including a variable containing the length of the interval and an indicator of death or survival on the interval. By using the log of the time on each interval as an offset, the model estimates the log of the mortality rate (rather than the count of deaths) as a linear function of the covariates. The resulting model provides mortality rate ratios (MRR) for the estimated parameters, which can be interpreted as relative risks of death.

I fit two sets of models for each of 4 outcomes, for 8 models in total. The outcomes I examined were: death by any natural cause, death by cancer, death by CVD, and death from all other (non-cancer/non-CVD) natural causes. The first set of models used a single term to differentiate lunar astronauts from all other astronauts. The other set of models used two binary terms to differentiate three groups of astronauts: those who orbited the Moon, those who walked on the Moon, and all non-lunar astronauts.

I assessed statistical significance and estimated confidence intervals around parameter estimates using standard errors generated from the robust sandwich variance estimator. This technique guards against violations of the assumption of the Poisson distribution that the mean is equal to the variance [36]. Significance testing on parameters was conducted at the $\alpha < 0.05$ level.

To assess how much bias may be present in the data, I checked covariate balance on 4 baseline covariates between lunar and non-lunar astronauts, using standardized mean differences (SMDs). Covariates with SMDs of less than 0.1 are considered to be in balance between two groups [37]. To test the statistical significance of the differences, I used t-tests for differences in continuous variables and chi-square tests with continuity correction for categorical variables. Imbalance and significant differences in baseline covariates may be an indicator of uncontrolled confounding in both measured and unmeasured variables [37].

3.2.4 Results

Table 1 displays demographic and actuarial characteristics of the lunar cohort of astronauts.

All of the lunar astronauts were selected to NASA between 1959 and 1966, all were male, and all were of White race/ethnicity. Of the 24 cohort members, 23 of them (96%) had a military background. Twelve (50%) of the astronauts had a Bachelor's degree, 10 (42%) had Master's degrees, and 2 (8%) had Doctoral degrees. The average age at first lunar mission was 39.0 years, and, as of 31 October 2018, the astronauts have been followed for an average of 41.8 years. **Figure 2** shows the breakdown of lunar astronaut deaths by lunar mission role and causal category.

Half of the lunar astronauts have died (12/24). Eight of these deaths were in the lunar surface group, and 4 in the lunar orbit group. Of the 12 total deaths, 3 of them were due to cancer, 3 were due to cardiovascular disease, and the remaining 5 were due to other natural causes. The single death due to external causes was experienced by a lunar surface astronaut.

Figure 3 provides a plot of the SMDs between the full lunar cohort and the rest of the astronaut corps for 4 baseline covariates. The figure shows that age, year of selection, education, and history of military service are all unbalanced between lunar and non-lunar astronauts, as each of their SMD values are greater than 0.1 (i.e., to the right of the red, vertical, dashed line). Statistical tests confirmed that the differences in these covariates are statistically significant, with p-values all less than 0.01 (values not pictured in **Figure 2**).

	All lunar astronauts (N = 24)		**Landed on the Moon (N = 12)**		**Orbited the Moon only (N = 12)**	
Characteristic	**n**	**(%)**	**n**	**(%)**	**n**	**(%)**
White	24	100.0	12	100.0	12	100
Males	24	100.0	12	100.0	12	100
Military experience	23	95.8	11	91.7	12	100
Education						
Bachelor	12	50.0	5	41.7	7	58.3
Master	10	41.7	5	41.7	5	41.7
PhD	2	8.3	2	8.3	0	0.0
	Mean	**(sd)**	**Mean**	**(sd)**	**Mean**	**(sd)**
Age, years						
Selection	32.8	1.9	32.6	2.2	33.0	1.6
First lunar mission	39.0	2.6	39.4	3.9	38.6	2.0
Death	73.2	13.5	78.7	9.4	64.3	16.4
End of study (survivors)	86.6	2.8	85.5	2.7	87.2	2.8
Follow-up time	41.3	11.9	41.6	10.0	41.0	14.1
Year of selection	1964	2.0	1963	2.1	1964	1.9
Year of lunar mission	1970	1.4	1970	1.4	1970	1.4

Table 1.
Lunar cohort characteristics.

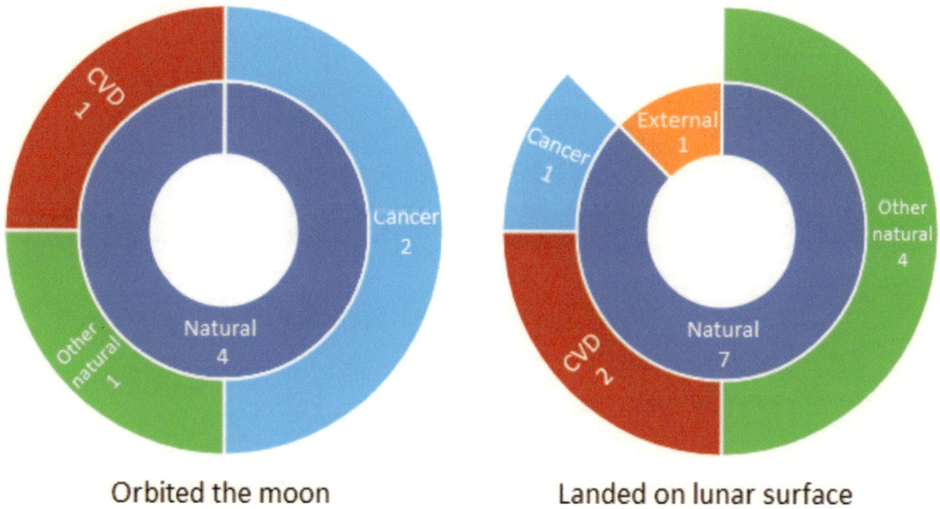

Figure 2.
Distribution of deaths by cause among lunar astronauts as of 31 October 2018.

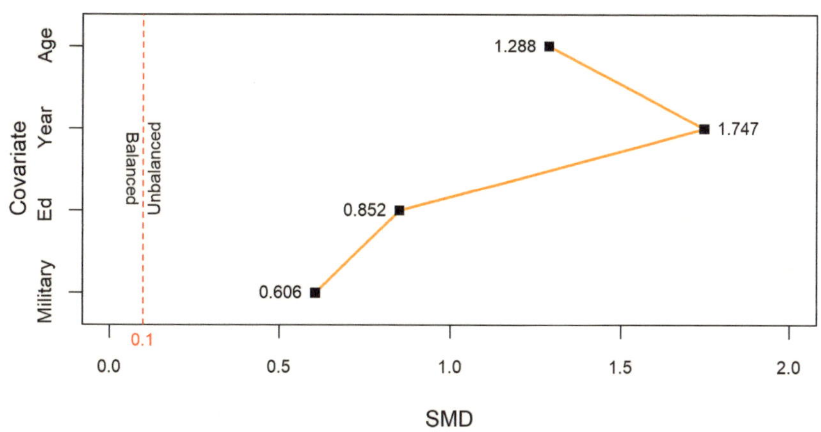

Figure 3.
Standardized mean differences (SMDs) for baseline covariates between lunar and non-lunar astronauts.

Figure 4 displays the MRR point estimates (blue squares) and 95% confidence intervals (blue lines), for the age, year terms, and the lunar covariate in each of the first four models. Looking at the point estimates for all covariates in all models, we see that none of them are statistically significant predictors of mortality rate. The estimates attached to age were all between a 5% and 22% increase in mortality rate per additional year past selection. Estimates for year of selection, calendar year, and being a lunar astronaut were all close to 1.0 with confidence intervals that included 1.0.

Figure 5 shows the MRRs for the second set of models, in which the lunar astronauts are further divided into orbital and surface subgroups. Once again age shows a consistent (though still statistically insignificant) increase in the mortality rate for all cause-of-death groups, with MRRs nearly identical to those in **Figure 4**. The point estimates for having orbited the Moon were less than 1.0 in all outcome models except cancer, although none of them were statistically significant. The largest estimated reduction in risk for astronauts who orbited the Moon was an MRR of 0.35 (95% CI = 0.03–3.52), for CVD.

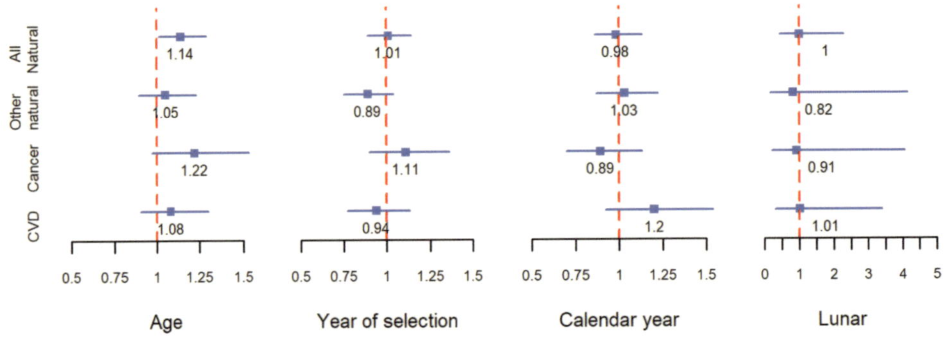

Figure 4.
Mortality rate ratios from Poisson models of mortality rates for all lunar astronauts.

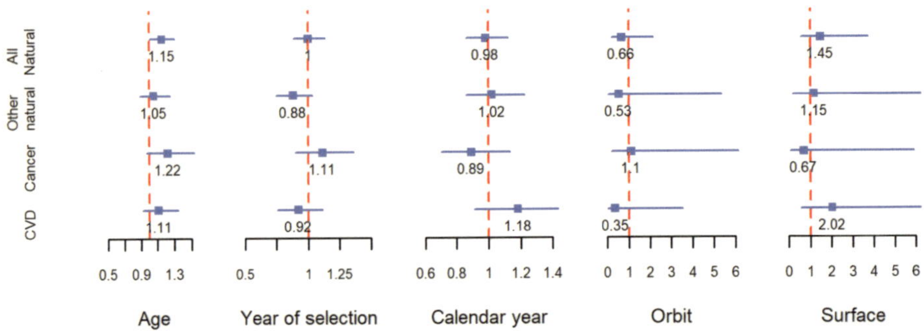

Figure 5.
Mortality rate ratios from Poisson models of mortality rates for orbital and surface lunar astronauts.

None of the MRRs for lunar surface astronauts were statistically significant. However, point estimates suggest approximately double the mortality risk from CVD (MRR = 2.02; 95% CI = 0.55–7.36), but about two-thirds the mortality risk from cancer (MRR = 0.67; 95% CI = 0.08–5.90). The net effect of these mixed results was an MRR for all natural causes of 1.45 (95% CI = 0.56–3.70).

3.2.5 Discussion

The models presented here fail to demonstrate any statistically significant increase in mortality risk for lunar astronauts in comparison to non-lunar astronauts. There are three potential reasons for this: (a) the models lack the statistical precision to discriminate excess risk from random variation; (b) the MRRs for lunar astronauts are confounded and thus misleading; or (c) lunar astronauts bear no excess mortality risk in comparison to other astronauts.

In regards to statistical precision of the models, small numbers of astronauts suggest, and prior studies have demonstrated poor statistical power in analyses of astronaut health [14, 15]. One study showed that adequate power was achieved for cancer mortality outcomes only when relative risks exceeded 4.0 for astronauts with radiation doses above the median [14]. Yet another study of 73 astronauts (including all 24 lunar astronauts) reported that, under reasonable assumptions about excess risk, statistical power was less than 6% [15]. Furthermore, power did not markedly improve even when assuming 10 times the number of observed deaths or 10 times the dose-dependent excess relative risk [15]. As statistical power is a function of sample size, the power of an even smaller subgroup analysis (such as lunar astronauts) would be lower still. Thus, it is entirely possible that the MRRs observed here are genuine differences, but that low statistical power alone means we cannot reliably distinguish them from random variation.

Another consequence of small sample size is the limited ability to adjust for other potential confounding variables in the models of **Figures 4** and 5, even when such variables are available. Because of this, the MRR estimates presented here may be confounded by other unmeasured factors that are not accounted for in the causal or statistical models. If so, this confounding could mask significant differences in mortality risk between lunar and non-lunar astronauts. **Figure 3** provides evidence of differences between lunar and non-lunar astronauts on at least 4 measured covariates, which may be evidence of confounding. Furthermore, the causal diagram of **Figure 1** suggests more unobserved confounding is possible. Of particular concern are unmeasured lifestyle factors such as tobacco use, alcohol use, diet, and exercise.

Though it is possible that low statistical precision or confounding is obscuring increased risk for lunar astronauts, it nevertheless seems implausible given the small exposure the lunar astronauts had to the major hazards of the lunar environment – hypogravity, cosmic radiation, and lunar dust. Specifically:

a. The total time of exposure to hypogravity for lunar astronauts pales in comparison to astronauts who have completed multiple shuttle missions or long-duration stays on the ISS. The mean time in space for lunar astronauts was 18.7 days, whereas the mean time in space for non-lunar astronauts (among those who have been to space) is 76.8 days. The average mission time for lunar astronauts was 8.7 days accrued on an average of 2 missions, while the average mission time for non-lunar astronauts was 26.1 days accrued on an average of 3 missions each. So, if exposure to hypogravity creates excess mortality risk – either in relation to total exposure time or unique number of exposures – non-lunar astronauts should be at greater risk than lunar astronauts. This is especially true if 0 G is more deleterious than lunar gravity, since the entirety of hypogravity exposure for non-lunar LEO astronauts has been in 0 G.

b. The doses of ionizing space radiation received by lunar astronauts are not extreme among astronauts or in comparison to Earth-based radiation workers. For example, as of 1993, lifetime cosmic radiation doses received by all astronauts, lunar included, were noted to be smaller than those received from medical diagnostic doses, [38] while a more recent report notes that only a third of the astronauts selected before 1970 received total space radiation doses greater than 11 mGy [33]. It should be noted that with the doses of space radiation received by all astronauts to date – even among those at the highest end of absorbed dose – that no acute effects have been observed, and only mildly increased risk of minor late effects has been observed [39, 40].

   Furthermore, the doses received by lunar astronauts would have to exceed those of non-lunar astronauts in order for lunar astronauts to be observed as at greater radiation-induced risk of mortality. Radiation dose is highly correlated with time in space, so just as non-lunar astronauts have logged considerably more time in space, they also have larger doses of cosmic radiation. Were cosmic radiation doses in the astronaut cohort as a whole at harmful levels we might therefore expect to see lunar astronauts at *reduced* mortality risk in comparison to non-lunar astronauts.

c. Though lunar dust is known to be more toxic than most dust found in industrial settings on earth, the astronauts' exposure to it was limited in both intensity and duration. Exposure to lunar dust was isolated, with direct contact limited to the time at rest in the lunar module and the trip back to earth in the command and service modules. All told this would equate to mere days of exposure at most. Workers on Earth who develop pneumoconiosis and other respiratory diseases typically only do so after repeated, sustained exposure to dust [26]. Based on what is known about dust exposure for workers on Earth, such brief exposure to lunar dust would therefore be unlikely to cause any lasting health problems, and thus unlikely to be a source of increased mortality for lunar astronauts.

In short, the exposures faced by lunar astronauts, though almost certainly harmful in sufficient quantity, were likely of inadequate intensity and duration to cause long-term excess mortality risk on historic missions.

4. Conclusions

The lunar environment poses a number of health risks for human explorers, some exotic and some familiar. Chief among them are hypobaric environments, hypogravity, cosmic radiation, and lunar dust. All of these, in sufficient doses, have the potential to cause a number of deleterious health effects.

The risk of hypobaric injury and death from the lack of atmosphere is omnipresent for astronauts living and working on the Moon. Deconditioning of the cardiovascular system in microgravity is a well-known hazard which is likely only partially abated by the Moon's low-gravity environment. Without appropriate protection, the radiation environment on the Moon is capable of producing an array of both acute and chronic health problems. Finally, lunar dust, as a highly fragmented and highly reactive substance, has the potential to cause respiratory illness, which may be exacerbated by differential patterns of particle deposition in the lungs under hypogravity conditions.

There have been only 24 individuals who have thus far visited the Moon, and to date they have shown no evidence of increased mortality rates for having done so. Yet, because of the small size of the cohort, it is possible that moderate increases in mortality risk from some causes may be in play. It is also possible that mortality estimates are confounded by uncontrolled factors. However, it seems unlikely that lunar astronauts are subject to excess mortality risk in comparison to non-lunar astronauts as a result of lunar missions. The duration of exposure to hypogravity and dose of cosmic radiation were low in comparison to non-lunar astronauts. Lunar astronauts' exposure to lunar dust was episodic and likely below the exposure threshold needed to trigger respiratory disease and thus increased mortality risk.

Though humanity currently has little direct experience with the lunar environment, our knowledge will continue to grow as humans return to the Moon in the decades to come. What we learn about how to successfully colonize the Moon will teach us valuable lessons that will allow us to colonize Mars and beyond. It thus becomes critical to continue to study human health in extraterrestrial environments, learning what we can as we can. The small steps of such ongoing efforts will thus enable the giant leaps that follow.

Acknowledgements

The author wishes to acknowledge Steven Day, PhD for helpful suggestions and edits that substantially improved this chapter. Also, the brave men and women who continue to explore space, and in so doing, further advance human knowledge.

Conflict of interest

The author has no conflict of interest to report.

Author details

Robert J. Reynolds
Mortality Research & Consulting, Inc., City of Industry, USA

*Address all correspondence to: rreynolds@mortalityresearch.com

References

[1] National Aeronautics and Space Administration. Moon fact sheet. Available at: https://nssdc.gsfc.nasa.gov/planetary/factsheet/moonfact.html [Last Accessed: 10-01-2019]

[2] Norfleet WT. Decompression-related disorders: Decompression sickness, arterial gas embolism, and ebullism syndrome. In: Barratt MR, Pool SL, editors. Principles of Clinical Medicine for Spaceflight. New York: Springer; 2008

[3] Billings CE. Barometric pressure. In: Parker JF, West VR, editors. Bioastronautics Data Book. 2nd ed. Washington D.C.: National Aeronautics and Space Administration; 1973

[4] Clement G. Fundamentals of Space Medicine. 2nd ed. New York: Springer; 2014

[5] Hamilton DR. Cardiovascular disorders. In: Barratt MR, Pool SL, editors. Principles of Clinical Medicine for Spaceflight. New York: Springer; 2008

[6] National Aeronautics and Space Administration. Risk of cardiac rhythm problems during spaceflight. Houston: Lyndon B. Johnson Space Center. Available at: https://humanresearchroadmap.nasa.gov/evidence/reports/Arrhythmia.pdf [Last Accessed: 10-01-2019]

[7] National Aeronautics and Space Administration. "HRR—Risk—Risk of Cardiac Rhythm Problems". Available at: https://humanresearchroadmap.nasa.gov/Risks/risk.aspx?i=79 [Last Accessed: 10-01-2019]

[8] Nicogossian AE, Doarn CR, Hu Y. Evolution of human capabilities and space medicine. In: Nicogossian et al., editors. Space Physiology and Medicine: From Evidence to Practice. 4th ed. New York: Springer; 2016

[9] Richter C, Braunstein B, Winnard A, Nasser M, Weber T. Human biomechanical and cardiopulmonary responses to partial gravity—A systematic review. Frontiers in Physiology. 2017;**8**:583

[10] Chancellor JC, Scott GB, Sutton JP. Space radiation: The number one risk to astronaut health beyond low earth orbit. Life (Basel). 2014;**4**(3):491-510

[11] Jones JA, Karouia F. Radiation disorders. In: Barratt MR, Pool SL, editors. Principles of Clinical Medicine for Spaceflight. New York: Springer; 2008

[12] Lockwood M, Hapgood M. The rough guide to the Moon and Mars. Astronomy & Geophysics. 2007;**48**(6):6.11-6.17

[13] Huff J et al. Risk of Radiation Carcinogenesis. NASA Human Research Program, Available at: https://humanresearchroadmap.nasa.gov/Risks/risk.aspx?i=96 [Last Accessed: 10-01-2019]

[14] Reynolds RJ. Development of a location exposure matrix for ionizing radiation in extraterrestrial environments and its application in the study of mortality for U.S. astronauts. Texas Medical Center Dissertations (via ProQuest). Paper AAI3604553

[15] Elgart SR et al. Radiation exposure and mortality from cardiovascular disease and Cancer in early NASA astronauts. Scientific Reports. 2018;**8**(1):8480. DOI: 10.1038/s41598-018-25467-9

[16] Patel Z et al. Risk of Cardiovascular Disease and Other Degenerative Tissue Effects from Radiation Exposure. NASA Human Research Program. Available at: https://humanresearchroadmap.nasa.gov/Risks/risk.aspx?i=98 [Last Accessed: 10-01-2019]

[17] Ade CJ, Broxterman RM, Charvat JM, Barstow TJ. Incidence rate of cardiovascular disease end points in the National Aeronautics and Space Administration astronaut corps. Journal of the American Heart Association. 2017;**6**:e005564. DOI: 10.1161/JAHA.117.005564

[18] McKay DS, Heiken G, Basu A, Blanford G, Simon S, Reedy R, et al. The lunar regolith. In: Heiken G, Vaniman D, French B, editors. Lunar Sourcebook. Cambridge: Cambridge University Press; 1991

[19] Greenberg PS, Chen D-R, Smith SA. Aerosol measurements of the fine and ultrafine particle content of lunar regolith. NASA Technical Reports Server, 2007. Available at: https://ntrs.nasa.gov/archive/nasa/casi.ntrs.nasa.gov/20070031906.pdf

[20] Cain JR. Lunar dust: The Hazard and astronaut exposure risks. Earth, Moon, and Planets. 2010;**107**:107-125

[21] McKay DS, Cooper BL, Taylor LA, James JT, Thomas-Keprta K, Pieters CM, et al. Physicochemical properties of respirable-size lunar dust. Acta Astronautica. 2015;**107**:163-176

[22] Grün E, Horanyi M, Sternovsky Z. The lunar dust environment. Planetary and Space Science. 2011;**59**:1672-1680

[23] Darquenne C, Prisk GK. Deposition of inhaled particles in the human lung is more peripheral in lunar than in normal gravity. European Journal of Applied Physiology. 2008;**103**:687-695

[24] Darquenne C, Prisk GK. Particulate deposition in the human lung under lunar habitat conditions. Aviation, Space, and Environmental Medicine. 2013;**84**:190-195

[25] Darquenne C, Borja MG, Oakes JM, Breen EC, Olfert IM, Scadeng M, et al. Increase in relative deposition of fine particles in the rat lung periphery in the absence of gravity. Journal of Applied Physiology. 2014;**117**:880-886

[26] Leung CC, Yu ITS, Chen W. Silicosis. Lancet. 2012;**379**:2008-2018

[27] James JT, Lam C, Scully RR. Comparative Benchmark Dose Modeling as a Tool to Make the First Estimate of Safe Human Exposure Levels to Lunar Dust. NASA Technical Reports Server, 2103. Available at: https://ntrs.nasa.gov/search.jsp?R=20130012803 [Last Accessed: 10-01-2019]

[28] Scully RR, Lam CW, James JT. Estimating safe human exposure levels for lunar dust using benchmark dose modeling of data from inhalation studies in rats. Inhalation Toxicology. 2013;**25**(14):785-793

[29] Lam C, Scully RR, Zhang Y, Renne RA, Hunter RL, McCluskey RA, et al. Toxicity of lunar dust assessed in inhalation-exposed rats. Inhalation Toxicology. 2013;**25**(12):661-678

[30] United States Department of Labor, Occupational Safety and Health Administration. OSHA Occupational Chemical Database. Silica, fused, respirable dust. Available at: https://www.osha.gov/chemicaldata/chemResult.html?recNo=442 [Last Accessed: 10-01-2019]

[31] National Aeronautics and Space Administration. Human Research Roadmap. Available at: https://humanresearchroadmap.nasa.gov/Risks/ [Last Accessed: 10-01-2019]

[32] Delp MD, Charvat JM, Limoli CL, Globus RK, Ghosh P. Apollo lunar astronauts show higher cardiovascular disease mortality: Possible deep space radiation effects on the vascular endothelium. Scientific Reports. 2016;**6**:29901

[33] Cucinotta FA, Hamada N, Little MP. No evidence for an increase in circulatory disease mortality in astronauts following space radiation exposures. Life Sciences and Space Research. 2016;**10**:53-56

[34] Reynolds RJ, Day SM. Mortality due to cardiovascular disease among Apollo lunar astronauts. Aerospace Medicine and Human Performance. 2017;**88**(5):492-496

[35] Jones M, Fowler R. Immortal time bias in observational studies of time-to event outcomes. Journal of Critical Care. 2016;**36**:195-199

[36] Zou G. A modified poisson regression approach to prospective studies with binary data. American Journal of Epidemiology. 2004;**159**(7):702-706

[37] Austin PC. The use of propensity score methods with survival or time-to-event outcomes: Reporting measures of effect similar to those used in randomized experiments. Statistics in Medicine. 2014;**33**:1242-1258

[38] Peterson LE, Pepper LJ, Hamm PB, Gilbert SL. Longitudinal study of astronaut health: Mortality in the years 1959-1991. Radiation Research. 1993;**133**(2):257-264

[39] Chylack LT Jr, Peterson LE, Feiveson AH, Wear ML, Manuel FK, Tung WH, et al. NASA study of cataract in astronauts (NASCA). Report 1: Cross-sectional study of the relationship of exposure to space radiation and risk of lens opacity. Radiation Research. 2009;**172**(1):10-20

[40] Chylack LT Jr, Feiveson AH, Peterson LE, Tung WH, Wear ML, Marak LJ, et al. NASCA report 2: Longitudinal study of relationship of exposure to space radiation and risk of lens opacity. Radiation Research. 2012;**178**(1):25-32